Analytical Modeling of Heterogeneous Cellular Networks

This self-contained introduction shows how stochastic geometry techniques can be used for studying the behavior of heterogeneous cellular networks (HCNs). The unified treatment of analytic results and approaches, collected for the first time in a single volume, includes the mathematical tools and techniques used to derive them. A single canonical problem formulation encompassing the analytic derivation of the signal to interference plus noise ratio (SINR) distribution in the most widely used deployment scenarios is presented, together with applications to systems based on the 3GPP- LTE standard, and with implications of these analyses on the design of HCNs. An outline of the different releases of the LTE standard and the features relevant to HCNs is also provided.

The book is a valuable reference for industry practitioners looking to improve the speed and efficiency of their network design and optimization workflow, and for graduate students and researchers seeking tractable analytical results for performance metrics in wireless HCNs.

Sayandev Mukherjee is a Senior Research Engineer at DOCOMO Innovations Inc., in Palo Alto, CA. He has worked at Bell Laboratories, Marvell Semiconductor Inc., and SpiderCloud Wireless Inc. Dr. Mukherjee has over seventy publications in journals and conferences, and has been awarded thirteen patents. He has been a Senior Member of the IEEE since 2005. He won the Wiley Best Paper Award at the International Workshop on Wireless Ad-hoc Networks (IWWAN) 2005 in London, UK.

Analytical Modeling of Heterogeneous Cellular Networks

Geometry, Coverage, and Capacity

SAYANDEV MUKHERJEE
DOCOMO Innovations Inc., Palo Alto, CA

CAMBRIDGE
UNIVERSITY PRESS

CAMBRIDGE
UNIVERSITY PRESS

University Printing House, Cambridge CB2 8BS, United Kingdom

One Liberty Plaza, 20th Floor, New York, NY 10006, USA

477 Williamstown Road, Port Melbourne, VIC 3207, Australia

314-321, 3rd Floor, Plot 3, Splendor Forum, Jasola District Centre, New Delhi - 110025, India

79 Anson Road, #06-04/06, Singapore 079906

Cambridge University Press is part of the University of Cambridge.

It furthers the University's mission by disseminating knowledge in the pursuit of
education, learning and research at the highest international levels of excellence.

www.cambridge.org
Information on this title: www.cambridge.org/9781107050945

First published 2014

A catalogue record for this publication is available from the British Library

Library of Congress Cataloging in Publication data
Mukherjee, Sayandev, 1970-
Analytical modeling of heterogeneous cellular networks : geometry, coverage, and capacity / Sayandev Mukherjee, DOCOMO Innovations Inc, Palo Alto, California.
pages cm
ISBN 978-1-107-05094-5 (hardback)
1. Cell phone systems. 2. Internetworking (Telecommunication) I. Title.
TK5103.2.M84 2014
621.3845′6–dc23 2013036071

ISBN 978-1-107-05094-5 Hardback
ISBN 978-1-107-67655-8 Paperback

Contents

Preface

The ever-rising demand for wireless data means that conventional cellular architectures based on large "macro" cells will soon be unable to support the anticipated density of high-data-rate users. Thus future wireless network standards envisaged by standards bodies like the Third Generation Partnership Project (3GPP), such as LTE Release 12 and later, rely on the following three ways of increasing system capacity: (a) additional spectrum; (b) enhanced spectral efficiency; and (c) offloading from the cellular network onto, say, WiFi.

So-called "small" cells are an attractive method of increasing spectral efficiency by means of spatial reuse of resources. Small cells can also exploit the fact that additional spectrum in the coming years will be freed up at higher frequencies, where path loss is higher than in the frequencies currently employed in macrocellular networks. A dense deployment of small cells can achieve the desired objective of high system capacity. However, such deployments are unlikely to be found outside of high-traffic areas such as major population centers. Thus, basic connectivity and mobility support will continue to be handled by macrocells. In other words, the wireless cellular network of the future is likely to be a *heterogeneous cellular network* (HCN), with more than one class of *base station* (BS).

How HCNs are studied and designed today

One of the most important metrics of network performance and user experience is the *signal to interference plus noise ratio*, or SINR, defined as the ratio of received signal power to the total received power from all sources other than the desired transmitter (i.e. from all *interferers*), plus the thermal noise power at the receiver (which is always present, even in the absence of any interferer). On a given link between a serving BS and a user, the SINR determines the bit error rate on the link, and therefore whether or not the user is *covered* (i.e. SINR exceeds some threshold), and, if it is covered, what the *capacity* (i.e. maximum achievable data rate) of that link is. Thus, a network operator wanting to optimize the operation of an HCN needs to know the *spatial distribution* of the SINR and its dependence on the deployment parameters of the HCN. On the downlink (i.e. a link for transmissions from a BS to a user), for example, this means the distribution of the SINR at an arbitrary user location in the HCN, and its dependence on the relative densities and transmit powers of the different classes, or *tiers*, of BSs in the network.

The usual method used to determine the spatial SINR distribution is extensive simulation. Simulation certainly has the advantage of being able to study any desired scenario to any desired depth of detail. However, this requires one to simulate separately every possible scenario of interest, including every possible choice of deployment parameters. As the number of combinations of deployment parameters rises exponentially in the number of tiers of the HCN, we see that HCN simulation scenarios are much more numerous than single-tier macrocellular network simulation scenarios, and an exhaustive simulation study of all possible scenarios of interest is time-consuming and expensive, if not altogether infeasible. Further, a partial investigation of a limited number of scenarios makes it difficult to draw inferences for new scenarios that have not been studied.

New results in modeling and analysis of HCNs

Since 2011, several theoretical results have contributed greatly to our understanding of the behavior of HCNs. We now have mathematically tractable analytical models, scalable to arbitrary numbers of tiers, that yield important insights into the SINR distribution throughout the network. Assuming the locations of BSs in the tiers are given by points of Poisson point processes (PPPs) in the plane, we can show that the *distribution of the SINR at an arbitrary location in the network can be calculated exactly and with low numerical complexity, and depends only on certain combinations of the network deployment parameters*. Given that, as recently as 2008, the analytic formula for the distribution of SINR in a single-tier network was unknown and this distribution had to be determined via simulation only, it is truly remarkable how far we have come in so short a time.

The objective of this book is to provide a self-contained exposition of the recent body of results, based on *stochastic geometry*, on SINR distribution in HCNs. Toward this goal, we also develop all the mathematical tools and techniques used to derive these results. It is seen that there is a remarkable unity in the problem of determining the SINR distribution, across all the different kinds of deployments of interest, and that the same basic set of techniques enables analytic treatment of all these scenarios. While it is the author's intention to sum up the "state of the art" as of the time of writing, this is currently a very active area of research, and new scenarios are being investigated every month. It is expected that the reader of this book will be well equipped to understand such future work, and even extend it.

A feasible and efficient workflow for an operator planning a new HCN deployment, or optimizing an existing one, is to use these analytic results to eliminate from consideration a large subset of possible deployments (specified by the deployment parameters) as unable to deliver the desired coverage and/or capacity. Then the few deployment parameter choices that are deemed interesting (based on this analysis) can be investigated exhaustively using simulation. Thus it is hoped that the book will also appeal to industry practitioners looking to apply these new results to improve network performance.

Outline of the book

This book focuses on the *downlink*, i.e. on links where the transmitter is one of the BSs in the system, and the receiver is a user. Thus when we talk of the distribution of SINR, we mean the distribution of SINR at the receiver. The following is a brief description of each chapter.

Introduction The case for the importance of HCNs, and the analytical investigation thereof: the importance of the SINR distribution in design and performance analysis, the infeasibility of studying every scenario via simulation alone, and the benefits of analytic modeling of BS and user locations in the network.

Structure of the SINR calculation problem In this chapter, we define the one mathematical problem that describes all the HCN scenarios that will be studied in the book. This chapter also introduces some key mathematical results from the theory of matrices that apply to the SINR distribution calculation problem. As an example of the application of this problem definition, we show mathematically why analytical treatment of the downlink SINR for the "classical" layout of BSs in a hexagonal lattice is impossible, but fairly straightforward if the BSs are points of a PPP.

Poisson point processes This chapter is intended as a self-contained "crash" course in the basic results from the theory of Poisson point processes (PPPs). The contents should be accessible to anyone with knowledge of calculus and do not require knowledge of measure theory. The most important results from the theory of homogeneous PPPs are easy to grasp intuitively. Rather than mathematically rigorous proofs, we offer heuristic arguments that should let the reader understand when to apply which result to the problem at hand.

SINR analysis for a single tier with fixed power This chapter and the following one comprise the main part of the book. They provide a full exposition of the recent advances in our understanding of the behavior of HCNs when the BS and user locations are modeled as points of independent PPPs. This chapter gives the main results for a single-tier network, so the results are applicable to the macrocellular networks of today.

SINR analysis for multiple tiers with fixed powers In this chapter, the results of the previous chapter are extended to apply to HCNs with multiple tiers. Special attention is given to analyzing several kinds of serving BS selection schemes. The joint distribution of the SINRs from candidate serving BSs, as well as the marginal SINR from the actual serving BS, are derived for different choices of candidate serving BSs from the tiers, and the overall serving BS. Applications include the study of camping and coverage probabilities in HCNs, and the need for inter-cell interference coordination (ICIC) when selection bias is applied across tiers for the selection of the overall serving BS from among the candidate serving BSs.

SINR analysis with power control In this chapter, we extend the ideas and results of the previous chapter to the case where the transmitters employ one of three types of power control: non-adaptive, open-loop, or closed-loop. In particular, we analyze the enhanced ICIC (eICIC) scheme in LTE macro-pico HCNs. We also study SINR distributions with a mix of transmitters employing fixed power and open-loop power control. Finally, we study a simple example of a network employing closed-loop power control and discuss some interesting aspects of overall system behavior.

Spectral and energy efficiency analysis Our results on SINR distributions translate directly to results on spectral efficiency of links to arbitrarily located users in the network. However, the area-averaged spectral efficiency in a cell is that averaged over transmissions to all the users in that cell served by a specific BS, and this is dependent upon the scheduling scheme employed by the BS. We mostly restrict ourselves to the simple round robin scheduler (RRS), but will also indicate how to analyze the spectral efficiency when the proportional-fair scheduler (PFS) is employed. We analyze a macro-pico HCN and show the spectral efficiency advantage of a standalone dense single-tier deployment. Finally, we show how to obtain an analysis of energy efficiency from the spectral efficiency.

Closing thoughts: future heterogeneous networks In this concluding chapter, we present some views on the face of future HCNs, and their key issues, challenges, and technologies. These include device-to-device (D2D) transmissions, cognitive radio, and the role of links conforming to the WiFi (IEEE 802.11) standard in future wireless cellular networks.

Intended readership of the book

This book tries to address two groups that overlap only partially:

(1) graduate students and researchers seeking tractable analytical results for topics in wireless HCNs;
(2) industry practitioners looking at improving the speed and efficiency of their network design and optimization workflow by moving away from the present exclusive reliance on simulations to a mix of analysis followed by simulation of a selected subset of scenarios deemed to be of interest from the analysis.

As already mentioned, the level of mathematical knowledge assumed for the reader is calculus and probability at the senior undergraduate level. A list of the probability distributions used in the book is provided in Appendix A. In particular, knowledge of measure theory is not required. Knowledge of 3GPP-LTE is not required, but familiarity with the LTE standard will help the reader to understand the applications of the results presented in the book. An outline of the different releases in the LTE standard and the features relevant to HCNs in each one is provided in Appendix B.

Acknowledgements

I would like to express my gratitude to Professor Jeffrey G. Andrews of the University of Texas, Austin, TX, for motivating my work in this area on three crucial occasions: first, with the seminal work in Andrews, Baccelli & Ganti (2011) for a single-tier network, which inspired me to extend it to Mukherjee (2011b) for multiple tiers; next, by urging me to submit a journal version of Mukherjee (2011b) to an upcoming special issue of the *IEEE JSAC*, which led to Mukherjee (2012a); and then by inviting me to give a talk at the 2012 Communication Theory Workshop, during the preparation of which I was led to the canonical probability formulation in Chapter 2.

I also owe a debt to my former colleague İsmail Güvenç, now Assistant Professor at Florida International University, Miami, FL, for inviting me to contribute a chapter to the book he was co-editing, which became Mukherjee (2013), and for introducing me to the topic of eICIC in the LTE standard, which led to our collaboration on Mukherjee & Güvenç (2011).

I would like to acknowledge the support for my research that I have received from my managers at DOCOMO Innovations Inc. (DII), Sean Ramprashad, Hiroyuki Ishii, and Fujio Watanabe, and in particular for the opportunity to apply stochastic geometry methods to analyze aspects of the DOCOMO Phantom Cell architecture proposal (Yu *et al.*, 2012; Mukherjee & Ishii, 2013), for which I owe a debt to Watanabe-san and Ishii-san.

If I had any doubts as to the feasibility of pulling off this book-writing project, they were put to rest by the enthusiasm with which my tentative proposal for the book was greeted by my director, Watanabe-san, my editor at Cambridge University Press, Phil Meyler, and by the DII general counsel, Jeremy Tucker, Esq., who negotiated the final contract with Cambridge. My heartfelt thanks to all of them.

There are as yet relatively few researchers applying stochastic geometry models to the analysis of wireless communication systems. It is a pleasure to acknowledge discussions with (in alphabetical order) Jeffrey Andrews, Francois Baccelli, Mehdi Bennis, Harpreet Dhillon, İsmail Güvenç, Martin Haenggi, Stephen Hanly, Robert Heath, Jr., Prasanna Madhusudhanan, Tony Quek, and Sundeep Rangan, among others. I also cherish the long discussions on this subject (and many others) that I had with my thesis advisor, Terrence L. Fine, now retired from Cornell University, Ithaca, NY, and with my DII colleagues Ulas Kozat, Haralabos (Babis) Papadopoulos, and Cedric Westphal (now at Huawei). A special thank you to Babis for the suggestion to use Map/Reduce to generate a list of keywords for the index.

Writing this book has involved many late nights, missed social engagements, and much time spent apart from my family. My wife Chandreyee has cheerfully borne the brunt of this, including single-handedly ferrying my children to all of their extracurricular activities. I cannot thank her enough for giving me the free time to complete the book. My children have motivated me in their own ways: my son Rik, 8, on hearing that I had a book contract, wrote (and bound) his own eight-page novel in three days and then accused me of slacking off in my own book-writing, while my daughter, Inika, then 5, made up for her lack of vocabulary by creating a picture book that had considerably more pages than her brother's. They make it all worthwhile, and this book is dedicated to them.

Notation

$\sim$	distributed as
${}_2F_1(a,b;c;z)$	Gaussian hypergeometric function
α	slope of path-loss model (path-loss exponent)
$\mathit{\Gamma}$	random variable denoting SINR
γ	SINR threshold
$\Gamma(z)$	gamma function
$\gamma(z,a)$	lower incomplete gamma function
Φ, Ψ	PPP of BS locations
$\tilde{\Phi}, \tilde{\Psi}$	PPP of received powers from the BSs of the PPP Φ or Ψ
x	scalar
$\boldsymbol{x}$	vector
$\mathbf{A}$	matrix
X	random variable
$f_X(x)$	PDF of X evaluated at x
$\boldsymbol{X}$	random vector
$f_{\boldsymbol{X}}(\boldsymbol{x})$	joint PDF of $\boldsymbol{X}$ evaluated at $\boldsymbol{x}$
j	$\sqrt{-1}$
K	intercept of path-loss model
b	BS belonging to a network
H_b	fade attenuation on link from BS b to user
Y_b	received power at user from BS b
n_{tier}	total number of tiers in the HCN
n_{open}	number of accessible (open) tiers in the HCN
I	serving tier
U_i	received power at user from candidate serving BS in tier i
V_i	total received power at user from all BSs in tier i other than the candidate serving BS
W	total received power at user from all BSs in HCN
$\mathcal{A}$	set or event
$1_{\mathcal{A}}(\cdot)$	indicator function of the set $\mathcal{A}$
$\mathbb{E}[X]$	expectation of the random variable X
$\mathbb{P}(\mathcal{A})$	probability of the event $\mathcal{A}$, also equal to $\mathbb{E}[1_{\mathcal{A}}]$
$\mathbb{R}$	the set of real numbers, also written $(-\infty,\infty)$
$\mathbb{R}_+$	the set of non-negative reals, also written $[0,\infty)$
$\mathbb{R}_{++}$	the set of positive reals, also written $(0,\infty)$

Acronyms and abbreviations

ABS	almost-blank subframe
BS	base station
(C)CDF	(complementary) cumulative distribution function
CDMA	code division multiple access
CLPC	closed-loop power control
CRE	cell range expansion
CSG	closed subscriber group
CSMA	carrier-sensed multiple access
CSR	complete spatial randomness
D2D	device-to-device
EM	expectation-maximization
FDD	frequency division duplexing
feICIC	further enhanced ICIC
HCN	heterogeneous cellular network
HeNB	home enhanced NodeB
ICIC	inter-cell interference coordination
ISD	inter-site distance
LTE	long-term evolution
LTE-A	LTE-Advanced
MHP	Matérn hard-core process
MIMO	multiple-input multiple-output
MMSE	minimum mean squared error
MTC	machine-type communication
OA	open access
OFDM	orthogonal frequency division multiplexing
OFDMA	orthogonal frequency division multiple access
OLPC	open-loop power control
P2P	peer-to-peer
PDF	probability density function
PFS	proportional-fair scheduler
PMF	probability mass function
PPP	Poisson point process
PSK	phase shift keying
REB	range expansion bias

RRH	remote radio head
RRS	round robin scheduler
RSRP	reference symbol received power
RSRQ	reference symbol received quality
SINR	signal to interference plus noise ratio
SIR	signal to interference ratio
TDD	time division duplexing
3GPP	Third Generation Partnership Project
TTI	transmit time interval
UE	user equipment
UMTS	universal mobile telecommunications system

1 Introduction

The rapid rate of increase in data traffic means that future wireless networks will have to support a large number of users with high data rates. A promising way to achieve this is by spectrum reuse through the deployment of cells with small range, such that the same time-frequency resources may be reused simultaneously in multiple cells. At the same time, the traditional *coverage* requirement for wireless users (supporting a modest rate at cell-edge users) is most economically met with cells having large range, i.e. the traditional *macrocellular* architecture. Thus the wireless cellular networks of the future are likely to be *heterogeneous*, i.e. have one or more *tiers* of small cells *overlaid* on the macrocellular tier.

Let us look at network design from the point of view of a service provider considering a deployment of a network in a certain region. Throughout this book, we only consider the *downlink*, i.e. the links from the BSs to the user terminals. The principal metrics we shall focus on are coverage and capacity.

(1) *Coverage* Intuitively, a user (or, more precisely, a user location) is *covered* if the communication link from the BS serving that user is sufficiently "good" that the user terminal can correctly receive both the control signaling and the data traffic at some minimum rate from the BS. Here, "good" means that (a) the received signal from the BS is "strong," i.e. the received signal power from the BS exceeds some threshold, and also that (b) the received *signal to interference plus noise ratio* (SINR) at the user exceeds some minimum value. In modern cellular networks such as LTE (3GPP 2010, Sec. 5.2.3.2), the threshold on received signal power is low enough to be comfortably exceeded by any modern receiver within a fairly large distance from the transmitting BS. Thus, for the purposes of analysis, the signal power requirement can be dropped without loss of generality, and the SINR is the key indicator of coverage.

(2) *Capacity* For the moment, the meaning of "capacity" is deliberately ambiguous, denoting both the maximum attainable data rate to a served user, and also the number of users that can be served by a given BS with some minimum data rate to each. We give more precise definitions of both coverage and capacity later, but for now it is sufficient to understand that the known relationship between achievable data rate on a link and the SINR on that link means that capacity in the sense used here is also dependent on SINR at the user terminals.

Since coverage and capacity are both related to the SINR at the user's receiver, we shall focus on the statistical properties of this SINR in the book. When we talk of the service provider *designing* a network to satisfy a certain coverage and/or capacity requirement, we mean the choice of deployment parameters for such a network, including:

(1) the number of tiers of the network;
(2) the densities of the BSs in the tiers; and
(3) the transmit powers of the BSs in the tiers.

Note that the service provider seeks to satisfy the coverage and/or capacity requirement with a deployment that is compatible with the service provider's targets for expected revenue, capital expenditure (CapEx), and operating expenditure (OpEx). In other words, there is a complicated *utility function* that depends on economic variables such as the pricing model for service, CapEx, OpEx, and expected revenue, in addition to the coverage criterion.

In this book, we focus only on the relationship between the deployment parameters of the network and the coverage and capacity. As we shall show, there are many sets of deployment parameters that are equivalent from the point of view of the SINR distribution. Thus, once an operating SINR distribution has been chosen, the service provider can choose one set of deployment parameters from the corresponding equivalence class based on its utility function, which takes into account the economic considerations.

When thermal noise is negligible, the SINR becomes the *signal to interference ratio*, or SIR. Before we can study the distribution of the SINR or SIR at a user, we need to define the wireless-channel model we shall be using in this book.

1.1 Wireless-channel model

There are two components to the so-called *link loss*, i.e. the difference between transmitted and received powers on a given wireless link: the distance-dependent *path loss* (which has no random component, and is completely known if the distance between transmitter and receiver is known), and the *fading* (which is random). We now discuss the models for these two components.

1.1.1 Path-loss model

We adopt the popular *slope-intercept model*, which says that the *path loss* (i.e. the ratio of the power transmitted over the link by the transmitter to the power received at the receiver) on the link between a BS in a given tier and an arbitrarily located user is given by

$$10 \log_{10} \frac{\text{power transmitted (by transmitter)}}{\text{power received (at receiver)}} = 10\alpha \log_{10}(\text{distance in meters}) - 10 \log_{10} K, \tag{1.1}$$

where the *slope* $\alpha > 2$ is the *path-loss exponent* and the *intercept* K is a constant accounting for the differences in height of the transmit and receive antennas, etc. It is important to bear in mind that α and K both depend upon the frequency band of operation.

Note that the path-loss model only applies to the attenuation over the air, i.e. after the transmission has left the transmit antenna, and just before entering the receive antenna. The final expression for received power can account for factors such as antenna gains as decrements to the above path loss.

1.1.2 Fading model

We show in the subsequent sections that the calculation of both the joint and marginal *complementary CDFs* (CCDFs) of the downlink SINR at an arbitrarily located user in the network are examples of a class of problems that are made tractable by pairing the PPP model for BS locations with a particular class of *independent identically distributed* (i.i.d.) fading processes on all links, namely a mixture of Nakagami-m (Mallik, 2010) fading processes with positive integer-valued m (see Appendix A for details on this and other distributions used in the book). The convention in wireless communications is to refer to the fading process by the name of the distribution (e.g. Rayleigh, Nakagami, etc.) of the *magnitude* of the equivalent complex baseband signal. The *fade attenuation* on a link is the square of this magnitude. For the Nakagami mixture fading process, the distributions of the fade attenuations on the links are given by a mixture of *Erlang* random variables, i.e. a mixture of gamma random variables with positive integer-valued shape parameters (see Appendix A). Moreover, the *probability density function* (PDF) of an arbitrary positive-valued continuous random variable can be approximated uniformly to any desired accuracy by a mixture of Erlang PDFs (see Lemma 2.6). The distribution of received power on a wireless fading channel with gamma-distribution fast fading and lognormal shadowing can be modeled by a single gamma random variable (of non-integer order), based on matching the first and second moments of the two distributions, as shown, for example, in Heath, Kountouris & Bai (2013). A similar approach (Atapattu, Tellambura & Jiang, 2011) shows how the distribution of received power in many wireless fading channel with lognormal shadowing can be approximated using mixtures of gamma PDFs (with non-integer order). Finally, Almhana *et al.* (2006, eqns. (16)–(17)) shows how to approximate such mixtures of general gamma PDFs by mixtures of Erlang PDFs.

Note that although it is customary in the literature to assume lognormal shadow fading on wireless links, we do not work with the lognormal distribution in our analysis. The reason is that the lognormal distribution is just not very tractable, which explains the plethora of approximations in the literature. Further, it has been observed, at least in a single-tier network, that lognormal shadowing has almost no effect on the (reciprocal of the) SIR (Błaszczyszyn, Karray & Klepper, 2010) when the serving BS is that received most strongly at the user.

The simplest example of a mixture of Erlang PDFs is the exponential PDF, corresponding to the Rayleigh fading process. We shall show that the joint or marginal

SINR CCDF for an arbitrary mixture of Erlang PDFs can be derived from the result for Rayleigh fading. For this reason, we begin by assuming *independent identically distributed* (i.i.d.) Rayleigh fading on all links from all BSs to any user location. Recall that, here, "Rayleigh" refers to the distribution of the magnitude of the equivalent complex baseband signal at the receiver. The fade attenuation (i.e. the power "gain" on the link to the receiver) is distributed as the square of a Rayleigh random variable, i.e. it is an exponential random variable (see Appendix A).

1.2 Distribution of the SINR at an arbitrary user

In this section, we shall make our first attempt at calculating, analytically, the distribution of the SINR at an arbitrary user location in the network. We begin with a snapshot of the wireless network *at a particular moment in time*. Further, let us assume there is just one tier of BSs, as in the currently deployed macrocellular networks, and that each BS transmits with fixed power, assumed to be the same at all BSs. Consider an arbitrarily located user, and suppose it is located at a distance R_0 from its serving BS (where the serving BS was selected according to some criterion – we discuss such criteria later). BSs in the network other than the serving BS are then *interferers*. Suppose there are M such interfering BSs, labeled $1, \ldots, M$, with their distances from the user being $R_1, \ldots, R_M$, respectively. Further, suppose that all BSs transmit with the same power P^{tx}, and that the fades on all links between the $M+1$ BSs (the serving BS, labeled 0, and the M interfering BSs $1, \ldots, M$) and the user are i.i.d. Rayleigh. Then the received power at the user from BS k ($k = 0, \ldots, M$) is given by

$$Y_k = \frac{PH_k}{R_k^\alpha}, \; k = 0, \ldots, M, \quad \{H_k\}_{k=0}^M \text{ i.i.d. Exp}(1), \quad P = KP^{\text{tx}}, \tag{1.2}$$

where the exponential distribution for H_k, $k = 0, \ldots, M$, follows from the i.i.d. Rayleigh fading assumption, and K and α are the intercept and slope, respectively, of the slope-intercept path-loss model (1.1) describing links from the BSs to the user. Let us further simplify by assuming that thermal noise power at the user's receiver is negligible compared to interference power. Thus the SINR (actually SIR) at the user is given by

$$\Gamma = \frac{Y_0}{\sum_{k=1}^M Y_k} = \frac{H_0/R_0^\alpha}{\sum_{k=1}^M H_k/R_k^\alpha}. \tag{1.3}$$

Conditioned on $R_k = r_k$, $k = 0, \ldots, M$, the CCDF of the SIR at the user can be written as follows (see Problem 1.1):

$$\mathbb{P}\{\Gamma > \gamma \mid R_0 = r_0, \ldots, R_M = r_M\} = \prod_{k=1}^M \mathbb{E}\left[\exp\left(-\gamma r_0^\alpha \frac{H_k}{r_k^\alpha}\right)\right]. \tag{1.4}$$

It is important to recognize that the conditional CCDF of the SINR is a *Laplace transform* of a sum of independent random variables:

$$\mathbb{P}\{\Gamma > \gamma \mid R_0 = r_0, \ldots, R_M = r_M\} = \mathbb{E}\, e^{-sW}\Big|_{s=\gamma r_0^{\alpha}}, \quad W = \sum_{k=1}^{M} \frac{H_k}{r_k^{\alpha}}. \tag{1.5}$$

Remark 1.1 The above calculation can be extended (see Problem 1.2) to the case where the common distribution of $H_0, \ldots, H_M$ is not exponential but Erlang (Torrieri & Valenti, 2012), i.e. the fading process is not Rayleigh but Nakagami-m with integer-valued m (Mallik, 2010). As we shall show later (see Lemma 2.6), this is sufficient to compute the conditional CCDF of the SINR to any desired degree of accuracy for any *arbitrary* fading process.

Remark 1.2 If M is infinite, the random variable W defined in (1.5) exists if and only if $\sum_{k=1}^{\infty} 1/r_k^{\alpha} < \infty$ (Haenggi & Ganti, 2009, p. 8).

Unfortunately, (1.5) is not quite the analytic *coup* it may seem. The reason is that (1.4) is the CCDF of the SIR at the user *conditioned* on the location of the user relative to the BSs $0, 1, \ldots, M$. In other words, (1.4) is the CCDF of the SIR at *one specific* user location. This is not very useful to a network operator, who wants to know either (a) the distribution of the SIR at an *arbitrary* user location, which is the *unconditional* CCDF of the SIR, or, equivalently, the expectation of (1.4) with respect to the joint distribution of $(R_0, R_1, \ldots, R_M)$; or (b) the CCDF of the SIR at an arbitrary user location at a distance of $R_0 = r_0$ from its serving BS, which is the expectation of (1.4) with respect to the joint distribution of $(R_1, \ldots, R_M)$ given $R_0 = r_0$. Now, for a *fixed* deployment of BSs in a network and an arbitrary user location in that network, analytic expressions for the joint distribution of $(R_0, R_1, \ldots, R_M)$ and the conditional joint distribution of $(R_1, \ldots, R_M)$ given $R_0 = r_0$ are unknown. Thus the expectation of (1.4) with respect to either of these distributions cannot be computed analytically.

Problems

1.1 Derive (1.4). *Hint*: Use (1.3) and the fact that $H_0 \sim \text{Exp}(1)$.

1.2 Show that (1.4) can be generalized to the case where $H_0, H_1, \ldots, H_M$ are i.i.d. Erlang(m, c) with shape parameter $m \in \{1, 2, \ldots\}$ and rate parameter $c > 0$:

$$f_{H_k}(x) = \frac{c^m}{(m-1)!} x^{m-1} \exp(-cx), \quad x \geq 0, \quad k = 0, 1, \ldots, M.$$

Hint: Observe that

$$f_{H_k}(x) = \frac{(-1)^{m-1}}{(m-1)!} c^m \frac{\partial^{m-1}}{\partial c^{m-1}} \exp(-cx), \quad x \geq 0,$$

i.e. the Erlang PDF is the $(m-1)$th derivative of the PDF of the Exp$(1/c)$ distribution with respect to the parameter c. Thus the Laplace transform of the Erlang PDF is given by the $(m-1)$th derivative with respect to c of the Laplace transform of the Exp$(1/c)$ PDF.

1.3 Why SINR distributions are usually found via simulation

For a given deployment of BSs, although the expectation of (1.4) with respect to either the joint distribution of $(R_0, R_1, \ldots, R_M)$ or the conditional joint distribution of $(R_1, \ldots, R_M)$ given $R_0 = r_0$ cannot be computed analytically, it can always be determined numerically via Monte Carlo methods. However, this is equivalent to *simulating* the network deployment. Further, a simulation of the network can be done on a much more detailed, much less simplified model of the wireless channel and its impairments while still keeping the computational complexity within reasonable bounds. This is the reason why simulation is so popular among industry practitioners – in addition to being just about the sole method to obtain these SIR distributions, it also allows one to probe the sensitivity of this SIR distribution to any parameter of the deployment or wireless-channel model in any desired depth of detail.

Moreover, for a single-tier macrocellular network with "regular" placement of BSs in the centers of hexagonal "cells" on the familiar hexagonal lattice, and for the usual values of path-loss exponents for macrocellular wireless links, it has been observed that the total downlink interference power at an arbitrary user location is essentially that due to the two "rings" of BSs (6 BSs in the first ring, 12 BSs in the second ring) around the cell containing the user. This permits the simulation to be efficient with regard to memory requirements and run time by restricting the deployment region to the "19-cell wraparound region" (Chen *et al.*, 2011), where *wraparound* means that the upper and lower, and the right and left, boundaries of the region are assumed contiguous, so there are no edge effects in the simulation, and statistics on the SINR may be drawn from all locations in the region without introducing bias.

It is important, however, to recall that the 19-cell wraparound model for simulation was proposed for the study of macrocellular networks. Let us examine the assumptions behind this model.

(1) *Regular placement of BSs* While this assumption was never applicable to any real-world deployment, it could at least be claimed to be more or less accurate for macrocells, because of careful planning of cell sites by operators. However, it is increasingly inapplicable as the number of cells increases and their size decreases, as is the case for future networks. The reason for this is that careful planning of cell sites is often impossible if the number of sites is large, and the deployment map then takes on a more "random" appearance.

(2) *Interference limited to two rings of BSs* With the placement of BSs no longer regular, the interference is effectively that due to all BSs within some distance from the user, instead of those BSs located in certain "rings." However, the value of this effective range depends upon the parameters of the path-loss model and the transmit powers of the BSs, which are different for the smaller cell sizes of future networks. In particular, for a macro-femto overlay network, these ranges are different for the two tiers, and a suitably sized region would need to be considered such that edge effects due to both tiers can be eliminated. This is likely to enlarge the deployment region to be simulated, increase memory requirements, and reduce the speed of the simulator.

To summarize, in future heterogeneous networks with smaller cell sizes, where the BS locations are more irregular, the simulations need to be done over larger deployment regions and will become more time-consuming.

A single-tier network has only two deployment parameters: the transmit powers of the BSs and their density (number of BSs per unit area). Thus the total number of scenarios required to be simulated for a single-tier network is not large, especially since the choices for transmit power and density of the BSs required for system operation are tightly constrained. However, each additional tier in the network multiplies the number of possibilities for the overall combination of transmit powers and densities of the various tiers. In other words, if our goal is exhaustive simulation of all feasible operating scenarios for a multi-tier heterogeneous network, the number of scenarios to be simulated rises exponentially with the number of tiers. If one of the tiers is that of user-owned (not operator-owned) femtocells, the number of scenarios is further augmented by the fact that some femtocells may be switched on and off by the user depending on the need for coverage at that user's premises. In other words, the layout of BSs may even become dynamic. Clearly, an exhaustive simulation-based investigation of all possible operating scenarios for a multi-tier heterogeneous network is challenging at best, and infeasible at worst.

1.4 The role of analytic modeling

The analytic-modeling-based investigation of deployment scenarios has two phases. In the first phase, we use probabilistic models for the locations of the BSs to determine analytic expressions for the CCDF of the SINR in the deployment region. In other words, the use of a stochastic model (Poisson point process, or PPP) for the locations of the BSs allows us to write an analytic expression for the expectation of (1.4) with respect to either the joint distribution of $(R_0, R_1, \ldots, R_M)$ or the conditional joint distribution of $(R_1, \ldots, R_M)$ given $R_0 = r_0$. Further, these results can be extended to arbitrary fading distributions and arbitrary numbers of tiers of BSs.

As we shall see, this has the benefit of providing insights into the *combinations* of deployment parameters that affect the CCDF of the SINR, and therefore the different sets of deployment parameters that are *equivalent* in that they yield the same CCDF of the SINR. This analytic phase allows us to sift through the large space of combinations of deployment parameters to settle quickly on certain equivalence classes of deployment parameters, each class corresponding to some desired CCDF of the SINR. The service provider may then choose a set of deployment parameters from one of these equivalence classes based on its economic utility function.

In the next phase of the network design, the shortlist of deployment scenarios (as defined by the deployment parameters) chosen in the first phase may be investigated in depth via simulation. This effectively uses the power of detailed simulation, incorporating all relevant aspects whose behavior and impact on performance is to be investigated, for a few selected deployment scenarios.

2 Structure of the SINR calculation problem

We begin with the simple SINR calculation problem of Chapter 1 and generalize it to a deployment with more than one tier of BSs. We then examine the features of the problem that permit us to evaluate the CCDF of the SIR in terms of Laplace transforms of fading coefficients on the links to the user location from the BSs in the tiers. Then we abstract the problem formulation slightly in order to define a general probability calculation of a vector of random variables, which we call the *canonical problem*. We derive general results and conditions under which the canonical probability may be expressed in terms of Laplace transforms of certain random variables. This is one half of the mathematical core of the book. The other half is the study of stochastic models that yield tractable analytic expressions for the Laplace transforms in the canonical probability calculation, and we discuss that topic in Chapter 3.

2.1 Statement of the SINR calculation problem

Let us return to the problem of computing the distribution of the SINR at an arbitrarily located user somewhere in a network with one or more tiers of BSs. We begin by defining the *candidate serving BSs*, the criteria for their selection, the criterion for choosing the BS that will serve the user, and some notation to represent received power from the candidate serving BSs and from all interferers.

2.1.1 Candidate serving BSs and the serving BS

Consider a snapshot of the wireless network at a particular moment in time. This corresponds to the model for a single resource element (a single subcarrier over one transmission interval) in the long-term evolution (LTE) standard, or to a single transmission interval for a frequency non-selective channel in the HSPA standard. Consider a user located anywhere in this network. Label the tiers of the network $1, \ldots, n_{\text{tier}}$, and assume that the user is only allowed to access the BSs in tiers $1, \ldots, n_{\text{open}}$. For example, a macro-femto HCN with a mix of open access (OA) and closed subscriber group (CSG) femtocells would be represented as an HCN with $n_{\text{tier}} = 3$ and $n_{\text{open}} = 2$, with tier 1 representing the macrocells, tier 2 the OA femtocells, and tier 3 the CSG femtocells.

Next, we assume that each BS in a tier transmits with the maximum power allowed for BSs in that tier. This immediately models the case of reference symbols (LTE) or pilot channels (HSPA), but also covers the case of data channels if we assume that the cells are all fully loaded. For any given user, a single *candidate serving BS* is chosen from each tier according to some criterion, and the BS that actually serves the user is chosen from among these candidate serving BSs according to some criterion. In Chapter 5, we study some commonly used criteria for selection of candidate serving BSs from the tiers, and for the overall serving BS from among the candidate serving BSs.

2.1.2 Basic definitions

Let Φ_i denote the set of locations of the BSs in tier i, $i = 1, \ldots, n_{\text{tier}}$. A BS in tier i is identified by its location $b \in \Phi_i$. In other words, we often write "the BS $b \in \Phi_i$" to mean "the BS in tier i located at b." Note that, in this notation, $b \in \mathbb{R}^2$.

Instantaneous received power at the user

A BS $b \in \Phi_i$ at distance R_b from the user transmits with power P_i^{tx} and is received at the user with power

$$(\forall b \in \Phi_i)\ Y_b = \frac{P_i H_b}{R_b^{\alpha_i}}, \quad \{H_b\}_{b \in \Phi_i} \text{ i.i.d.}, \quad P_i = K_i P_i^{\text{tx}}, \quad i = 1, \ldots, n_{\text{tier}}, \tag{2.1}$$

where α_i and K_i are, respectively, the slope and intercept of the path-loss model (1.1) that describes links between BSs in tier i and the user location, and we assume that the fading processes on all links from all BSs to the user are *independent*. The fade attenuations $\{H_b\}_{b \in \Phi}$ on the links to the user from the BSs in tier i are therefore independent. Further, the fade attenuations on the links to the user from all BSs in tier i (except possibly from the candidate serving BS in tier i, if $i \in \{1, \ldots, n_{\text{open}}\}$ – see below) are assumed *identically distributed* with common *continuous* distribution described by the PDF $f_{H_i}(\cdot)$. Finally, let us assume that the thermal noise power at the user receiver (measured over the same bandwidth as the received powers) is N_0.

For any tier $i = 1, \ldots, n_{\text{tier}}$, the total received power at the user from all BSs in tier i is given by

$$W_i \equiv \sum_{b \in \Phi_i} Y_b = \sum_{b \in \Phi_i} \frac{H_b}{R_b^{\alpha_i}}, \quad i = 1, \ldots, n_{\text{tier}}.$$

For each *open* or *accessible* tier $i = 1, \ldots, n_{\text{open}}$, let us denote the *candidate serving* BS from tier i by B_i. The received power at the user from the candidate serving BS in accessible tier i is therefore $U_i \equiv Y_{B_i}$, while the interference from all other BSs in the accessible tier i is

$$V_i \equiv \sum_{b \in \Phi_i \setminus \{B_i\}} Y_b = \sum_{b \in \Phi_i \setminus \{B_i\}} \frac{H_b}{R_b^{\alpha_i}} = W_i - U_i, \quad U_i \equiv Y_{B_i}, \quad i = 1, \ldots, n_{\text{open}}. \tag{2.2}$$

Serving tier, serving BS, and instantaneous SINR

If there are multiple accessible tiers in the network, one of the *candidate serving* BSs is chosen (based on any of several criteria to be defined later) as the *serving* BS. If there is only one accessible tier, the candidate serving BS for this tier (chosen according to some criterion) is also the serving BS. Let us denote the index of the *serving tier* by I. Then I is a random variable taking values in $\{1,\ldots,n_{\text{open}}\}$, and the serving BS is B_I. If $I=i$, the SINR at the user is given by

$$\Gamma_i \equiv \frac{U_i}{\sum_{\substack{j=1\\ j\neq i}}^{n_{\text{open}}} U_j + \left[\sum_{k=1}^{n_{\text{open}}} V_k + \sum_{k=n_{\text{open}}+1}^{n_{\text{tier}}} W_k + N_0\right]}, \quad i=1,\ldots,n_{\text{open}}. \tag{2.3}$$

Special case: single tier

Consider a deployment with a single tier: $n_{\text{tier}} = n_{\text{open}} = 1$. Then $I=1$ with probability 1. Without loss of generality, denote the distance of the serving BS B_1 from the user by R_*, and the corresponding fading attenuation by H_* (instead of H_{B_1}). Dropping the subscript $i=1$ for convenience, the SINR at the user is given from (2.3) by

$$\Gamma = \frac{U}{V+N_0}, \quad U = \frac{PH_*}{R_*^\alpha}, \quad P = KP^{\text{tx}}, \tag{2.4}$$

and V is given by (2.2) with $i=1$. Suppose now that the common distribution of all the fading attenuations on all links to the user location from all BSs (including H_*, on the link from the serving BS to the user) is the exponential distribution with unit mean, Exp(1). It follows that the CCDF of Γ is given by

$$\mathbb{P}\{\Gamma > \gamma \,|\, R_* = r_*\} = \exp\left(-\frac{\gamma r_*^\alpha}{P} N_0\right) \mathbb{E}\left[\exp\left(-\frac{\gamma r_*^\alpha}{P} V\right) \,\Big|\, R_* = r_*\right], \tag{2.5}$$

which may be seen as the counterpart of the probability (1.4), defined by conditioning only on the distance to the serving BS.

2.2 SINR distributions

The *cumulative distribution function* (CDF) of the wideband SINR at an arbitrary user is often referred to as the *geometry* in the literature. It is a representation of the "environment" that an arbitrarily located user in the network will encounter.

Throughout the book, we are interested in two kinds of joint SINR CCDFs, which we define below.

2.2.1 Joint CCDF of SINRs from candidate serving BSs

The first is the joint CCDF of the SINRs at the user when receiving from the candidate serving BSs of (all or a subset of) the open tiers:

$$\mathbb{P}\{\Gamma_{j_1} > \gamma_1, \ldots, \Gamma_{j_k} > \gamma_k\}, \quad \gamma_1 > 0, \ldots, \gamma_k > 0, \tag{2.6}$$

for all $k \le n_{\text{open}}$ and distinct indices $(j_1,\ldots,j_k)$ such that $1 \le j_1 < j_2 < \cdots < j_k \le n_{\text{open}}$.

Remark 2.1 The *inclusion-exclusion formula* (see Problem 2.1) lets us write the probability of the union of k sets $\mathcal{A}_1, \ldots, \mathcal{A}_k$ in terms of all possible distinct intersections of $m \leq k$ of these sets:

$$\mathbb{P}\left(\bigcup_{l=1}^{k} \mathcal{A}_l\right) = \sum_{m=1}^{k} (-1)^{m-1} \sum_{\substack{l_1,\ldots,l_m: \\ 1 \leq l_1 < \cdots < l_m \leq k}} \mathbb{P}\left(\mathcal{A}_{l_1} \cap \cdots \cap \mathcal{A}_{l_m}\right). \tag{2.7}$$

Defining $\mathcal{A}_l \equiv \{\Gamma_{j_l} \leq \gamma_l\}$, $l = 1, \ldots, k$, from (2.7), we can obtain the joint PDF of the SINR from the joint CCDF (2.6) as follows (see Problem 2.2):

$$\begin{aligned} &f_{\Gamma_{j_1},\ldots,\Gamma_{j_k}}(\gamma_1, \ldots, \gamma_k) \\ &= \frac{(-1)^k \partial^k}{\partial \gamma_1 \cdots \partial \gamma_k} \mathbb{P}\{\Gamma_{j_1} > \gamma_1, \ldots, \Gamma_{j_k} > \gamma_k\}, \; [\gamma_1, \ldots, \gamma_k]^\top \in \mathbb{R}_{++}^k, \end{aligned} \tag{2.8}$$

so for any $\mathcal{A} \subseteq \mathbb{R}^k$, $\mathbb{P}\{(\Gamma_{j_1}, \ldots, \Gamma_{j_k}) \in \mathcal{A}\}$ can be computed by integrating the joint PDF (2.8) over $\mathcal{A}$. However, if $\mathcal{A}$ is of the form $(\underline{\gamma}_1, \overline{\gamma}_1] \times \cdots \times (\underline{\gamma}_k, \overline{\gamma}_k]$, then (see Problem 2.3)

$$\mathbb{P}\{(\Gamma_{j_1}, \ldots, \Gamma_{j_k}) \in \mathcal{A}\} = \mathbb{P}\{\underline{\gamma}_l < \Gamma_{j_l} \leq \overline{\gamma}_l, \; l = 1, \ldots, k\}$$

can be written directly in terms of sums and differences of probabilities of the form (2.6).

2.2.2 Joint CCDF of SINRs from BSs ordered by serving BS selection criterion

We stated previously that the serving BS selection criterion yields the serving BS, either directly in the case of a single-tier deployment, or through the serving tier index I if there are multiple tiers. In fact, the serving BS selection criterion *ranks* the candidate serving BSs (if there are multiple tiers) or just the BSs of the deployment (if there is only one tier) in some order, as "best," "second-best," etc., and then selects the "best" one as the serving BS. In addition to the marginal CCDF of the SINR at the user when receiving from the serving (i.e. the "best") BS, we may also be interested in the joint CCDF of the SINRs at the user when receiving from, say, the serving ("best") and "second-best" BSs. The knowledge of this joint distribution enables us to perform analyses of coverage and, more importantly, spectral efficiency, under coordinated transmission schemes, where the user may receive transmissions from both the "best" and "second-best" BSs, or may be served by the "second-best" BS if the "best" BS is overloaded.

In other words, we are also interested in a second kind of joint SINR CCDF,

$$\mathbb{P}\{\Gamma_{(1)} > \gamma_1, \ldots, \Gamma_{(k)} > \gamma_k\}, \quad \gamma_1 > 0, \ldots, \gamma_k > 0 \tag{2.9}$$

for all $k \leq n_{\text{open}}$, where the SINRs of interest, denoted $\Gamma_{(1)}, \ldots, \Gamma_{(k)}$ are those at the user when receiving from the "best" k candidate serving BSs (if there are multiple tiers) or just BSs (if there is only one tier), when these "best" k candidate serving BSs are arranged in decreasing order as determined by the serving BS selection criterion.

Note that $\Gamma_{(1)} \equiv \Gamma_I$ by definition, so, if $k = 1$, (2.9) yields the marginal CCDF of the SINR at the user when receiving from the serving BS:

$$\mathbb{P}\{\Gamma_I > \gamma\}, \quad \gamma > 0.$$

Remark 2.2 We study several criteria for serving BS selection in Chapter 5. A widely used approach in serving tier selection is to apply *selection bias* to load different tiers of a multi-tier network preferentially. Then, owing to selection bias in a multi-tier deployment, it is not true in general that $\Gamma_{(1)} > \Gamma_{(2)} > \cdots > \Gamma_{(n_{\text{open}})}$, even if the candidate serving BS in each tier is the one that is received most strongly at the user. In other words, for any $i = 1, \ldots, k$, $\Gamma_{(i)}$ is, by definition, the SINR at the user when receiving from the overall ith "best" candidate serving BS, but it is not necessarily true that $\Gamma_{(i)}$ is the overall ith largest SINR.

Finally, we note that *the distribution of the SIR can always be obtained from the distribution of the SINR simply by setting the thermal noise power N_0 to zero*.

2.2.3 Conventions and notation

Remark 2.3 It is important to emphasize that we impose the strict requirement $\gamma_l > 0$, and not the looser requirement $\gamma_l \geq 0$ in (2.6) and (2.9) for $l = 1, \ldots, k$. The reason is that if $\gamma_l = 0$ for some $l \in \{1, \ldots, k\}$, then, because Γ_{j_l} in (2.6) and $\Gamma_{(l)}$ in (2.9) exceed γ_l with probability 1, they simply drop out of the corresponding joint CCDFs, which therefore no longer describe the distribution of Γ_{j_l} and $\Gamma_{(l)}$, respectively. For example, in (2.6) if $\gamma_l = 0$ we have

$$\begin{aligned}
&\mathbb{P}\{\Gamma_{j_1} > \gamma_1, \ldots, \Gamma_{j_k} > \gamma_k\} \\
&\quad = \mathbb{P}\{\Gamma_{j_1} > \gamma_1, \ldots, \Gamma_{j_{l-1}} > \gamma_{l-1}, \Gamma_{j_{l+1}} > \gamma_{l+1}, \ldots, \Gamma_{j_k} > \gamma_k\},
\end{aligned}$$

which is not a descriptor of the distribution of Γ_{j_l}.

In the rest of the book, we use "positive-valued" in the strict sense, i.e. if $x \geq 0$, we say that x is non-negative-valued, whereas if $x > 0$ we say that x is positive-valued. We also employ the equivalent notation $x \in \mathbb{R}_+$ in place of $x \geq 0$, and $x \in \mathbb{R}_{++}$ in place of $x > 0$.

For any two vectors $\boldsymbol{v} = [v_1, \ldots, v_n]^\top$ and $\boldsymbol{u} = [u_1, \ldots, u_n]^\top$, define $\boldsymbol{v} > \boldsymbol{u} \Leftrightarrow v_l > u_l$, $l = 1, \ldots, n$, with similar definitions for $\boldsymbol{v} \geq \boldsymbol{u}$, $\boldsymbol{v} < \boldsymbol{u}$, and $\boldsymbol{v} \leq \boldsymbol{u}$. We also define

$$\mathbf{0}_n = [\underbrace{0, \ldots, 0}_{n}]^\top, \qquad \boldsymbol{e}_k^{(n)} = [\underbrace{0, \ldots, 0}_{k-1}, 1, \underbrace{0, \ldots, 0}_{n-k}]^\top, \quad k = 1, \ldots, n.$$

Occasionally, we omit the dimension subscript and simply write $\mathbf{0}$, as in $\boldsymbol{u} > \mathbf{0}$.

2.3 The canonical SINR probability

In this section, we define the SINR distributions of interest to us, and show that they all share a common structure, which we call the *canonical SINR probability*. Thus, being able to compute the canonical SINR probability will yield the desired distributions of the SINR.

2.3.1 Form of joint CCDF of SINRs from candidate serving BSs

Observe that for $[\gamma_1, \ldots, \gamma_k]^\top \in \mathbb{R}^k_{++}$, i.e. for $[\gamma_1, \ldots, \gamma_k]^\top > \mathbf{0}_k$, and using the notation of Section 2.1.2, the event $\Gamma_{j_1} > \gamma_1, \ldots, \Gamma_{j_k} > \gamma_k$ in (2.6) may be written as (see Problem 2.4)

$$\begin{bmatrix} 1-\rho_1 & -\rho_1 & \cdots & -\rho_1 \\ -\rho_2 & 1-\rho_2 & \cdots & -\rho_2 \\ \vdots & \cdots & \ddots & \vdots \\ -\rho_k & -\rho_k & \cdots & 1-\rho_k \end{bmatrix} \begin{bmatrix} U_{j_1} \\ U_{j_2} \\ \vdots \\ U_{j_k} \end{bmatrix} > W(j_1, \ldots, j_k) \begin{bmatrix} \rho_1 \\ \rho_2 \\ \vdots \\ \rho_k \end{bmatrix}, \tag{2.10}$$

where

$$\rho_1 = \frac{\gamma_1}{1+\gamma_1}, \rho_2 = \frac{\gamma_2}{1+\gamma_2}, \ldots, \rho_k = \frac{\gamma_k}{1+\gamma_k}$$

and

$$W(j_1, \ldots, j_k) = V_{j_1} + \cdots + V_{j_k} + \sum_{\substack{l=1 \\ l \notin \{j_1, \ldots, j_k\}}}^{n_{\text{tier}}} W_l + N_0. \tag{2.11}$$

In other words, the probability calculation (2.6), rewritten as (2.10), can be seen as *the calculation of a probability of the form*

$$\mathbb{P}\{\mathbf{A}\boldsymbol{X} > W\boldsymbol{\rho}\},$$

where $\boldsymbol{X} = [U_{j_1}, \ldots, U_{j_k}]^\top$, the square matrix $\mathbf{A}$ has all off-diagonal entries ≤ 0, and $\boldsymbol{\rho} = [\rho_1, \ldots, \rho_k]^\top$ is a constant vector with the property (imposed by the requirements in Remark 2.3) that all its entries are positive, i.e. $\boldsymbol{\rho} > \mathbf{0}$.

2.3.2 Form of joint CCDF of SINRs from BSs ordered by serving BS selection criterion

We do not discuss here the criteria used for serving BS selection from among the candidate serving BSs in a multi-tier deployment, leaving those details for Chapter 5. However, we shall state here that our principal interest is in serving BS selection criteria that employ the received powers from the candidate serving BSs, together with positive bias factors, one per tier, to *rank* the candidate serving BSs in decreasing order of $\tau_i Y_{B_i}$, $\tau_i > 0$, $i = 1, \ldots, n_{\text{open}}$. (Note that, as the same bias factor is used for all BSs in a tier, this form of biasing has no effect on serving BS selection in a single-tier deployment.) In other words, the serving BS selection criterion returns tier labels $I_1 (\equiv I), I_2, \ldots, I_{n_{\text{open}}}$ such that

$$\tau_{I_1} Y_{B_{I_1}} \geq \tau_{I_2} Y_{B_{I_2}} \geq \cdots \geq \tau_{I_{n_{\text{open}}}} Y_{B_{I_{n_{\text{open}}}}}.$$

Let us denote the SINRs from the candidate serving BSs of the tiers $I_1, \ldots, I_{n_{\text{open}}}$ by

$$\Gamma_{(i)} \equiv \Gamma_{I_i}, \quad i = 1, \ldots, n_{\text{open}}.$$

Then we can show (see (5.86)) that a joint CCDF of the form (2.9) can be written as sums and differences of terms of the form

$$\mathbb{P}\{\Gamma_{j_1} > \gamma_1', \ldots, \Gamma_{j_k} > \gamma_k', \ (\forall l \in \{k+1, \ldots, m\})\ U_{j_l} > (\tau_{j_k}/\tau_{j_l}) U_{j_k}\}, \qquad m = k, \ldots, n_{\text{open}}, \qquad (2.12)$$

where $\gamma_1', \ldots, \gamma_k'$ are functions of the parameters $\tau_{j_1}, \ldots, \tau_{j_k}$ and of the arguments $\gamma_1, \ldots, \gamma_k$ in (2.9). Following the same steps as in the derivation of (2.10), let us write (2.12) for each $i = 1, \ldots, n_{\text{open}}$ in the form (see (5.88))

$$\mathbb{P}\{\mathbf{A}\boldsymbol{X} > [W\boldsymbol{\rho}^\top, \mathbf{0}_{m-k}^\top]^\top\} = \mathbb{P}\{\mathbf{A}\boldsymbol{X} > W\boldsymbol{b}\}, \quad \boldsymbol{b} = [\boldsymbol{\rho}'^\top, \mathbf{0}_{m-k}^\top]^\top, \ \boldsymbol{\rho}' > \mathbf{0}_k,$$

where $\boldsymbol{X}(= [U_{j_1}, \ldots, U_{j_k}]^\top)$ and W are the same as before, the square matrix $\mathbf{A}$ again has all off-diagonal entries ≤ 0, but is now of dimension $m \times m$, and the only difference from the formulation in Section 2.3.1 is that now $\boldsymbol{b}$ may have zero-valued entries. Note that $\boldsymbol{b}$ cannot have all zero entries because $\boldsymbol{\rho}' > \mathbf{0}$, so $\boldsymbol{b} \geq \mathbf{0}$ but $\boldsymbol{b} \neq \mathbf{0}$.

2.3.3 Joint CCDF of SINR in canonical probability form

From the discussion in Sections 2.3.1 and 2.3.2, we see that, in order to obtain an analytic expression for either of the two kinds of joint CCDFs we are interested in, we need only calculate analytically a probability of the form

$$\mathbb{P}\{\mathbf{A}\boldsymbol{X} > W\boldsymbol{b}\}, \qquad (2.13)$$

where

(1) W and the entries of the vector $\boldsymbol{X}$ are random variables taking positive values with probability 1,
(2) the square matrix $\mathbf{A}$ has all off-diagonal entries ≤ 0, and
(3) $\boldsymbol{b} \in \mathbb{R}_+^k \setminus \{\mathbf{0}_k\}$ is a constant vector.

We further restrict ourselves to the case where $\boldsymbol{X}$ and W are independent, and call the resulting probability calculation problem the *canonical probability problem*. Note that, in the context of a heterogeneous network deployment, whether $\boldsymbol{X}$ and W are independent will depend upon the criterion for choosing the candidate serving BSs in the tiers. For example, this is true if the candidate serving BS in each tier is the one that is nearest to the user location, but not true if it is the one that is received most strongly at the user location. For now, we focus on the canonical probability problem as an abstract probability calculation without reference to the physical meaning of the random variables. Later, we show how to apply the results to several practically important HCN deployment scenarios, including that where the candidate serving BS in each tier is the BS from that tier that is received most strongly at the user location.

Problems

2.1 Prove (2.7). *Hint*: Start with the result for $k = 2$, $\mathbb{P}(\mathcal{A}_1 \cup \mathcal{A}_2) = \mathbb{P}(\mathcal{A}_1) + \mathbb{P}(\mathcal{A}_2) - \mathbb{P}(\mathcal{A}_1 \cap \mathcal{A}_2)$, then use mathematical induction for $k > 2$.

2.2 Derive (2.8). *Hint*: Begin by writing the complement of the joint CDF as follows:

$$1 - \mathbb{P}\{\Gamma_{j_1} \leq \gamma_{j1}, \ldots, \Gamma_{j_k} \leq \gamma_{jk}\} = \mathbb{P}\left(\{\Gamma_{j_1} > \gamma_{j1}\} \cup \cdots \cup \{\Gamma_{j_k} > \gamma_{jk}\}\right).$$

Then use the inclusion-exclusion formula (2.7) to write the probability on the right side as a sum of probabilities of intersections of l sets out of the collection $\{\Gamma_{j_1} > \gamma_{j1}\}, \ldots, \{\Gamma_{j_k} > \gamma_{jk}\}$ with signs $(-1)^{l-1}$, $l = 1, \ldots, k$. Finally, use the definition of the joint PDF as the kth mixed partial derivative of the joint CDF, and the fact that the kth mixed partial derivative of a function of $l < k$ of the variables must be zero.

2.3

(1) Verify that $\mathbb{P}\{\underline{\gamma}_{jl} < \Gamma_{j_l} \leq \overline{\gamma}_{jl}, l = 1, \ldots, k\}$ can be written in the form $\mathbb{P}\{W\underline{\boldsymbol{b}} < \mathbf{A}\boldsymbol{X} \leq W\overline{\boldsymbol{b}}\}$ for some $\mathbf{0}_k \leq \underline{\boldsymbol{b}} \leq \overline{\boldsymbol{b}}$.

(2) Prove that, for any n, any random vector $\boldsymbol{Y} = [Y_1, \ldots, Y_n]^\top$ and any $\underline{\boldsymbol{b}}, \overline{\boldsymbol{b}} \in \mathbb{R}^n$ such that $\underline{\boldsymbol{b}} < \overline{\boldsymbol{b}}$,

$$\mathbb{P}\{\underline{\boldsymbol{b}} < \boldsymbol{Y} \leq \overline{\boldsymbol{b}}\} = \sum_{\boldsymbol{\epsilon} \in \{0,1\}^n} (-1)^{\mathbf{1}_n^\top \boldsymbol{\epsilon}}\, \mathbb{P}\{\boldsymbol{Y} > [c_{1,1+\epsilon_1}, \ldots, c_{n,1+\epsilon_n}]^\top\},$$

where $\boldsymbol{\epsilon} = [\epsilon_1, \ldots, \epsilon_n]^\top$ and $\mathbf{C} = [\underline{\boldsymbol{b}}\ \overline{\boldsymbol{b}}] = [c_{i,j}]_{i=1,\ldots,n, j=1,2} \in \mathbb{R}^{n \times 2}$. *Hint*: If $\boldsymbol{Y}_{(1)} = [Y_2, \ldots, Y_n]^\top$, and similarly for $\underline{\boldsymbol{b}}_{(1)}$ and $\overline{\boldsymbol{b}}_{(1)}$, prove by integrating the joint PDF $f_{\boldsymbol{Y}}(\boldsymbol{y}) = f_{Y,\boldsymbol{Y}_{(1)}}(y, \boldsymbol{y}_{(1)})$ that

$$\begin{aligned}\mathbb{P}\{\underline{\boldsymbol{b}} < \boldsymbol{Y} \leq \overline{\boldsymbol{b}}\} &= \mathbb{P}\{Y_1 > \underline{b}_1, \underline{\boldsymbol{b}}_{(1)} < \boldsymbol{Y}_{(1)} \leq \overline{\boldsymbol{b}}_{(1)}\} \\ &\quad - \mathbb{P}\{Y_1 > \overline{b}_1, \underline{\boldsymbol{b}}_{(1)} < \boldsymbol{Y}_{(1)} \leq \overline{\boldsymbol{b}}_{(1)}\}.\end{aligned}$$

Then expand the probabilities involving $\boldsymbol{Y}_{(1)}$ similarly, and so on.

(3) Use the above result to show that $\mathbb{P}\{\underline{\gamma}_{jl} < \Gamma_{j_l} \leq \overline{\gamma}_{jl}, l = 1, \ldots, k\}$ can be evaluated using sums and differences of terms of the form $\mathbb{P}\{\mathbf{A}\boldsymbol{X} > W\boldsymbol{c}\}$.

2.4 Verify that (2.6) can be rewritten as (2.10). *Hint*: Apply the identity $a/b > c \Leftrightarrow a/(a+b) > c/(1+c)$ to rewrite each of the conditions $\Gamma_{j_l} > \gamma_l$, $l = 1, \ldots, k$.

2.4 Calculation of the canonical probability

2.4.1 Z-matrices and M-matrices

Recall that we want to compute in closed form the *canonical probability* $\mathbb{P}\{\mathbf{A}\boldsymbol{X} > W\boldsymbol{b}\}$, where $\boldsymbol{b} \geq \mathbf{0}$, $\boldsymbol{b} \neq \mathbf{0}$, all entries in $\boldsymbol{X}$, W are independent, and all off-diagonal entries of the square matrix $\mathbf{A}$ are ≤ 0. Such matrices are called *Z-matrices*. Let the dimension of $\boldsymbol{b}$ and $\boldsymbol{X}$ be n, so that the Z-matrix $\mathbf{A}$ is $n \times n$.

Given $\boldsymbol{u} \in \mathbb{R}^n$ and $\mathcal{V} \subset \mathbb{R}^n$, we also define $\boldsymbol{u} + \mathcal{V} = \{\boldsymbol{u} + \boldsymbol{v} : \boldsymbol{v} \in \mathcal{V}\}$. With this notation, we begin our analysis of the canonical probability problem with the following result.

LEMMA 2.1 *Given $\tilde{\boldsymbol{b}} \in \mathbb{R}^n_+ \setminus \{\mathbf{0}_n\}$ and an $n \times n$ Z-matrix $\mathbf{A}$ (i.e. every off-diagonal entry in $\mathbf{A}$ is ≤ 0), the set $\{\boldsymbol{x} \in \mathbb{R}^n_+ : \mathbf{A}\boldsymbol{x} > \tilde{\boldsymbol{b}}\}$ is non-empty if and only if $\mathbf{A}$ is an M-matrix*[1] *(i.e. $\mathbf{A}^{-1}$ exists, and every entry in $\mathbf{A}^{-1}$ is ≥ 0).*

Proof Suppose $\mathbf{A}$ is not an M-matrix. From (I_{28}) in Berman & Plemmons (1994, Ch. 6, Thm. 2.3, p. 136), it follows that, for all $\boldsymbol{x} \in \mathbb{R}^n$, if $\mathbf{A}\boldsymbol{x} > \mathbf{0}_n$ it cannot be true that $\boldsymbol{x} \in \mathbb{R}^n_+ \setminus \{\mathbf{0}_n\}$. Further, $\mathbf{0}_n \notin \{\boldsymbol{x} \in \mathbb{R}^n_+ : \mathbf{A}\boldsymbol{x} > \tilde{\boldsymbol{b}}\}$ because $\tilde{\boldsymbol{b}} \neq \mathbf{0}_n$. Thus

$$\{\boldsymbol{x} \in \mathbb{R}^n_+ : \mathbf{A}\boldsymbol{x} > \tilde{\boldsymbol{b}}\} = \{\boldsymbol{x} \in \mathbb{R}^n_+ \setminus \{\mathbf{0}_n\} : \mathbf{A}\boldsymbol{x} > \tilde{\boldsymbol{b}}\} \subseteq \{\boldsymbol{x} \in \mathbb{R}^n_+ \setminus \{\mathbf{0}_n\} : \mathbf{A}\boldsymbol{x} > \mathbf{0}_n\} = \emptyset.$$

On the other hand, if $\mathbf{A}$ is an M-matrix, then every entry of $\mathbf{A}^{-1}$ is non-negative and $\mathbf{A}^{-1}$ exists, so $\mathbf{A}^{-1}$ cannot have an all-zero row. In other words, $\mathbf{A}^{-1}\boldsymbol{\beta} > \mathbf{0}_n$ for all $\boldsymbol{\beta} > \mathbf{0}_n$. Therefore

$$\begin{aligned}\{\boldsymbol{x} \in \mathbb{R}^n_+ : \mathbf{A}\boldsymbol{x} > \tilde{\boldsymbol{b}}\} &= \mathbf{A}^{-1}\tilde{\boldsymbol{b}} + \{\boldsymbol{x} \in \mathbb{R}^n_+ : \mathbf{A}\boldsymbol{x} > \mathbf{0}_n\} \\ &= \mathbf{A}^{-1}\tilde{\boldsymbol{b}} + \{\boldsymbol{x} \in \mathbb{R}^n_+ : (\exists \boldsymbol{\beta} > \mathbf{0}_n)\, \mathbf{A}\boldsymbol{x} = \boldsymbol{\beta}\} \\ &= \mathbf{A}^{-1}\tilde{\boldsymbol{b}} + \{\mathbf{A}^{-1}\boldsymbol{\beta} : \boldsymbol{\beta} > \mathbf{0}_n\} \subseteq \mathbb{R}^n_{++},\end{aligned} \tag{2.14}$$

from which it follows that $\{\boldsymbol{x} \in \mathbb{R}^n_+ : \mathbf{A}\boldsymbol{x} > \tilde{\boldsymbol{b}}\}$ is non-empty and contains only vectors with all-positive entries. □

In Problem 2.5, we provide a direct proof of Lemma 2.1 when the Z-matrix $\mathbf{A}$ has the specific structure of the matrix on the left side of (2.10).

2.4.2 Expressions for $\mathbb{P}\{AX > \tilde{b}\}$

$X_1, \ldots, X_n$ are independent exponentially distributed random variables

From Lemma 2.1 we see that, if $X_1, \ldots, X_n$ are independent exponentially distributed random variables for any $n \geq 1$ and $\mathbf{A}$ is an $n \times n$ matrix, we can calculate in closed form the probability $\mathbb{P}\{\mathbf{A}\boldsymbol{X} > \tilde{\boldsymbol{b}}\}$ for any $\tilde{\boldsymbol{b}} \in \mathbb{R}^n_+ \setminus \{\mathbf{0}_n\}$.

LEMMA 2.2 *Suppose $X_1, \ldots, X_n$ are independent and $X_k \sim \text{Exp}(1/c_k)$ with $c_k > 0$, $k = 1, \ldots, n$. Define $\boldsymbol{c} = [c_1, \ldots, c_n]^\top$. Then, for any $\tilde{\boldsymbol{b}} \in \mathbb{R}^n_+ \setminus \{\mathbf{0}_n\}$ and any $n \times n$ Z-matrix $\mathbf{A}$, we have*

$$\mathbb{P}\{\mathbf{A}\boldsymbol{X} > \tilde{\boldsymbol{b}}\} = \begin{cases} 0, & \text{if } \mathbf{A} \text{ is not an M-matrix,} \\ \left(\prod_{k=1}^n c_k\right) \exp\left(-\boldsymbol{c}^\top \mathbf{A}^{-1}\tilde{\boldsymbol{b}}\right) h_{\mathbf{A}^{-1}}(\boldsymbol{c}), & \text{if } \mathbf{A} \text{ is an M-matrix,} \end{cases}$$

where, for any arbitrary matrix $\mathbf{B} \in \mathbb{R}^{n \times n}$ with all non-negative entries,

$$h_{\mathbf{B}}(\boldsymbol{c}) = \det \mathbf{B} \prod_{k=1}^n \frac{1}{\boldsymbol{c}^\top \mathbf{B} \boldsymbol{e}_k^{(n)}}, \quad \boldsymbol{c} \in \mathbb{R}^n_{++}. \tag{2.15}$$

Proof The result when $\mathbf{A}$ is not an M-matrix follows immediately from Lemma 2.1. If $\mathbf{A}$ is an M-matrix, we write $\mathbf{A}^{-1} = [\boldsymbol{a}_1^{(-1)}, \ldots, \boldsymbol{a}_n^{(-1)}]^\top$, where $\boldsymbol{a}_k^{(-1)} = \mathbf{A}^{-1}\boldsymbol{e}_k^{(n)} \in \mathbb{R}^n_+ \setminus \{\mathbf{0}_n\}$ is the kth column of $\mathbf{A}^{-1}$, $k = 1, \ldots, n$. Note that, for any $\boldsymbol{\beta} = [\beta_1, \ldots, \beta_n]^\top$,

[1] Some authors refer to this definition as that of a non-singular M-matrix.

$\mathbf{A}^{-1}\boldsymbol{\beta} = \sum_{k=1}^{n} \beta_k \boldsymbol{a}_k^{(-1)}$. The joint PDF of $\boldsymbol{X} = [X_1, \ldots, X_n]^\top$ is $f_{\boldsymbol{X}}(\boldsymbol{x}) = (\prod_{k=1}^{n} c_k)\exp(-\boldsymbol{c}^\top \boldsymbol{x})$, $\boldsymbol{x} \in \mathbb{R}_+^n$. Then from (2.14) we can write

$$
\begin{aligned}
\mathbb{P}\{\mathbf{A}\boldsymbol{X} > \tilde{\boldsymbol{b}}\} &= \int_{\{\boldsymbol{x}\in\mathbb{R}_+^n : \mathbf{A}\boldsymbol{x} > \tilde{\boldsymbol{b}}\}} \left(\prod_{k=1}^{n} c_k\right) \exp(-\boldsymbol{c}^\top \boldsymbol{x}) \mathrm{d}\boldsymbol{x} \\
&= \exp\left(-\boldsymbol{c}^\top \mathbf{A}^{-1}\tilde{\boldsymbol{b}}\right) \int_{\{\mathbf{A}^{-1}\boldsymbol{\beta} : \boldsymbol{\beta}\in\mathbb{R}_{++}^n\}} \left(\prod_{k=1}^{n} c_k\right) \exp(-\boldsymbol{c}^\top \boldsymbol{x}) \mathrm{d}\boldsymbol{x} \\
&= \exp\left(-\boldsymbol{c}^\top \mathbf{A}^{-1}\tilde{\boldsymbol{b}}\right) \det \mathbf{A}^{-1} \int_{\mathbb{R}_{++}^n} \left(\prod_{k=1}^{n} c_k\right) \exp\left\{-\sum_{k=1}^{n}\left[\boldsymbol{c}^\top \boldsymbol{a}_k^{(-1)}\right]\beta_k\right\} \mathrm{d}\boldsymbol{\beta} \\
&= \left(\prod_{k=1}^{n} c_k\right) \exp\left(-\boldsymbol{c}^\top \mathbf{A}^{-1}\tilde{\boldsymbol{b}}\right) \det \mathbf{A}^{-1} \prod_{k=1}^{n} \int_0^\infty \exp\left\{-\left[\boldsymbol{c}^\top \boldsymbol{a}_k^{(-1)}\right]\beta_k\right\} \mathrm{d}\beta_k \\
&= \left(\prod_{k=1}^{n} c_k\right) \exp\left(-\boldsymbol{c}^\top \mathbf{A}^{-1}\tilde{\boldsymbol{b}}\right) h_{\mathbf{A}^{-1}}(\boldsymbol{c}),
\end{aligned} \tag{2.16}
$$

which proves the result. □

$X_1, \ldots, X_n$ are independent hyper-Erlang distributed random variables

As in Problem 1.2, Lemma 2.2 can be generalized to the case where $X_1, \ldots, X_n$ are independent and each X_k has a so-called *hyper-Erlang* distribution whose PDF is a convex combination (i.e. a *mixture*) of Erlang PDFs:

$$
f_{X_k}(x) = \sum_{l=1}^{p_k} \beta_{k,l}\, g(x; m_{k,l}, c_{k,l}),\ x \geq 0, \quad \sum_{l=1}^{p_k} \beta_{k,l} = 1, \tag{2.17}
$$

where $p_k \in \{1, 2, \ldots\}$, $m_{k,l} \in \{0, 1, 2, \ldots\}$, $\beta_{k,l} > 0$, $c_{k,l} > 0$, $l = 1, \ldots, p_k$, $k = 1, \ldots, n$, and, for any $c > 0$ and $m \in \{0, 1, 2, \ldots\}$,

$$
g(x; m, c) = \frac{c^{m+1} x^m}{m!} \exp(-cx), \quad x \geq 0. \tag{2.18}
$$

LEMMA 2.3 *For the same hypotheses as Lemma 2.2 except that now X_k has the PDF (2.17), $k = 1, \ldots, n$, the canonical probability is given by*

$$
\mathbb{P}\{\mathbf{A}\boldsymbol{X} > \tilde{\boldsymbol{b}}\}
= \begin{cases}
0, & \mathbf{A} \text{ not an M-matrix}, \\
\displaystyle\sum_{l_1=1}^{p_1} \cdots \sum_{l_n=1}^{p_n} \left\{ \prod_{k=1}^{n} \frac{(-1)^{m_{k,l_k}} \beta_{k,l_k} (c_{k,l_k})^{m_{k,l_k}+1}}{m_{k,l_k}!} \right\} & \\
\quad \times \dfrac{\partial^{m_{1,l_1}+\cdots+m_{n,l_n}}}{\partial (c_{1,l_1})^{m_{1,l_1}} \cdots \partial (c_{n,l_n})^{m_{n,l_n}}} \left\{ h_{\mathbf{A}^{-1}}\left(\left[c_{1,l_1}, \ldots, c_{n,l_n}\right]^\top\right) \right. & \\
\quad \left. \times \exp\left(-\left[c_{1,l_1}, \ldots, c_{n,l_n}\right]^\top \mathbf{A}^{-1}\tilde{\boldsymbol{b}}\right) \right\}, & \mathbf{A} \text{ an M-matrix}.
\end{cases} \tag{2.19}
$$

Proof From the independence of $X_1, \ldots, X_n$, their joint PDF is given by

$$f_{\boldsymbol{X}}(\boldsymbol{x}) = \prod_{k=1}^{n} f_{X_k}(x_k) = \prod_{k=1}^{n} \sum_{l_k=1}^{p_k} \beta_{k,l_k} \frac{(c_{k,l_k})^{m_{k,l_k}+1} x_k^{m_{k,l_k}}}{m_{k,l_k}!} \exp\left(-c_{k,l_k} x_k\right)$$

$$= \sum_{l_1=1}^{p_1} \cdots \sum_{l_n=1}^{p_n} \left\{ \prod_{k=1}^{n} \beta_{k,l_k} \frac{(c_{k,l_k})^{m_{k,l_k}+1} x_k^{m_{k,l_k}}}{m_{k,l_k}!} \right\} \exp\left(-\left[c_{1,l_1}, \ldots, c_{n,l_n}\right]^\top \boldsymbol{x}\right)$$

$$= \sum_{l_1=1}^{p_1} \cdots \sum_{l_n=1}^{p_n} \left\{ \prod_{k=1}^{n} \frac{(-1)^{m_{k,l_k}} \beta_{k,l_k} (c_{k,l_k})^{m_{k,l_k}+1}}{m_{k,l_k}!} \right\}$$

$$\times \frac{\partial^{m_{1,l_1}+\cdots+m_{n,l_n}}}{\partial(c_{1,l_1})^{m_{1,l_1}} \cdots \partial(c_{n,l_n})^{m_{n,l_n}}} \exp\left(-\left[c_{1,l_1}, \ldots, c_{n,l_n}\right]^\top \boldsymbol{x}\right). \tag{2.20}$$

We can now redo the derivation of (2.16), with the quantity in curly braces in (2.20) taking the place of $\prod_{k=1}^{n} c_k$, and applying the mixed partial derivative in (2.20) to the integral over $\boldsymbol{x}$, to obtain (2.19). □

2.4.3 Expressions for the canonical probability $\mathbb{P}\{\mathbf{A}\boldsymbol{X} > W\boldsymbol{b}\}$

We are now ready to derive expressions for the canonical probability $\mathbb{P}\{\mathbf{A}\boldsymbol{X} > W\boldsymbol{b}\}$ for the calculation of the joint CCDF of the SINR.

$X_1, \ldots, X_n$ are independent exponentially distributed random variables

Now we set $\tilde{\boldsymbol{b}} = W\boldsymbol{b}$, where W is a random variable taking positive values with probability 1 and $\boldsymbol{b}$ is a constant. Then from Lemma 2.2 we have the following theorem.

THEOREM 2.4 *Suppose* $X_k \sim \text{Exp}(1/c_k)$ *with* $c_k > 0$, $k = 1, \ldots, n$; W *is a random variable taking positive values with probability* 1, *and* $W, X_1, \ldots, X_n$ *are all independent. Define* $\boldsymbol{X} = [X_1, \ldots, X_n]^\top$ *and* $\boldsymbol{c} = [c_1, \ldots, c_n]^\top$. *Then the canonical probability* $\mathbb{P}\{\mathbf{A}\boldsymbol{X} > W\boldsymbol{b}\}$ *for any* $\mathbf{0}_n \neq \boldsymbol{b} \geq \mathbf{0}_n$ *is given by*

$$\mathbb{P}\{\mathbf{A}\boldsymbol{X} > W\boldsymbol{b}\} = \mathbb{E}\left[\mathbb{P}\{\mathbf{A}\boldsymbol{X} > W\boldsymbol{b} \mid W\}\right]$$

$$= \begin{cases} 0, & \mathbf{A} \text{ not an M-matrix}, \\ \left(\prod_{k=1}^{n} c_k\right) h_{\mathbf{A}^{-1}}(\boldsymbol{c}) \mathcal{L}_W(\boldsymbol{c}^\top \mathbf{A}^{-1} \boldsymbol{b}), & \mathbf{A} \text{ an M-matrix}, \end{cases}$$

where $h_{\mathbf{A}^{-1}}(\boldsymbol{c})$ *is given by* (2.15), *and*

$$\mathcal{L}_W(s) = \mathbb{E}[\exp(-sW)], \quad s > 0,$$

is the Laplace transform *of (the PDF of) the random variable* W.

Thus we see that, if $W, X_1, \ldots, X_n$ are independent and $X_k \sim \text{Exp}(1/c_k)$, $k = 1, \ldots, n$, the canonical probability problem has a complete analytic solution. Further, the canonical probability is zero if $\mathbf{A}$ is not an M-matrix, whereas if $\mathbf{A}$ is an M-matrix *the canonical probability is given in terms of the Laplace transform of* W.

$X_1, \ldots, X_n$ are independent hyper-Erlang distributed random variables

A similar application of Lemma 2.3 provides us with our main result.

THEOREM 2.5 *For the same hypotheses as Theorem 2.4, except that now X_k has the PDF (2.17), $k = 1, \ldots, n$, the canonical probability is given by*

$$\mathbb{P}\{\mathbf{A}\boldsymbol{X} > W\boldsymbol{b}\}$$

$$= \begin{cases} 0, & \mathbf{A} \text{ not an } M\text{-matrix}, \\ \displaystyle\sum_{l_1=1}^{p_1} \cdots \sum_{l_n=1}^{p_n} \left\{ \prod_{k=1}^{n} \frac{(-1)^{m_{k,l_k}} \beta_{k,l_k} (c_{k,l_k})^{m_{k,l_k}+1}}{m_{k,l_k}!} \right\} & \\ \displaystyle\times \frac{\partial^{m_{1,l_1}+\cdots+m_{n,l_n}}}{\partial (c_{1,l_1})^{m_{1,l_1}} \cdots \partial (c_{n,l_n})^{m_{n,l_n}}} \left\{ h_{\mathbf{A}^{-1}} \left(\left[c_{1,l_1}, \ldots, c_{n,l_n}\right]^{\top} \right) \right. & \\ \left. \times \mathcal{L}_W \left(\left[c_{1,l_1}, \ldots, c_{n,l_n}\right]^{\top} \mathbf{A}^{-1} \boldsymbol{b} \right) \right\}, & \mathbf{A} \text{ an } M\text{-matrix}. \end{cases}$$

2.4.4 Approximating arbitrary PDFs by mixtures of Erlang PDFs

The following result shows that Theorem 2.5 is sufficient to calculate the canonical probability when $X_1, \ldots, X_n$ are independent and arbitrarily distributed, because any distribution of a non-negative-valued random variable can be uniformly approximated to any desired accuracy by a hyper-Erlang distribution.

LEMMA 2.6 *Let $f_X(x)$, $x \geq 0$, be the PDF of a non-negative-valued random variable X. Then, for any $\epsilon > 0$, there exists a positive integer N_ϵ, positive integers $m_1, \ldots, m_{N_\epsilon}$, positive $c_1, \ldots, c_{N_\epsilon}$, and non-negative $\beta_1, \ldots, \beta_{N_\epsilon}$ with $\sum_{l=1}^{N_\epsilon} \beta_l = 1$ such that, for any $x \geq 0$,*

$$\left| f_X(x) - \sum_{l=1}^{N_\epsilon} \beta_l \frac{c_l^{m_l}}{(m_l - 1)!} x^{m_l - 1} \exp(-c_l x) \right| < \epsilon.$$

Proof See DeVore & Lorentz (1993, Prob. 5.6, p. 14) or Kelly (2011, Exer. 3.3.3(i), p. 80). An interesting proof of convergence of the corresponding CDFs is contained in Williams (2001, (6.6.L5), p. 214). □

Remark 2.4 It turns out that the hyper-Erlang distribution is a member of a general class of distributions called *phase type distributions* which all have the property of being able to approximate any arbitrary distribution. An efficient use of the expectation-maximization (EM) algorithm to obtain a hyper-Erlang distribution to approximate a given distribution is described in Thummler, Buchholz & Telek (2006). For wireless fading channels, another EM-based algorithm is given in Almhana, Choulakian & McGorman (2006), but this approach starts with a hyper-gamma approximation which is then "rounded" to a hyper-Erlang approximation.

Problem

2.5 Prove Lemma 2.1 when $\mathbf{A}$ is the $k \times k$ Z-matrix on the left-hand side (LHS) of (2.10). *Hint*: Write the matrix as

$$\mathbf{A} = \mathbf{I}_k - \boldsymbol{\rho}\mathbf{1}_k^\top, \quad \boldsymbol{\rho} = [\rho_1, \ldots, \rho_k]^\top, \quad \mathbf{1}_k = [\underbrace{1, \ldots, 1}_{k}]^\top, \tag{2.21}$$

where $\mathbf{I}_k$ is the $k \times k$ identity matrix. Then proceed as follows.

(1) Prove that

$$\det(\mathbf{I}_k - \boldsymbol{\rho}\mathbf{1}_k^\top) = 1 - \mathbf{1}_k^\top \boldsymbol{\rho}. \tag{2.22}$$

Hint: Use the identity

$$\begin{bmatrix} \mathbf{I}_k & \mathbf{0}_k \\ \mathbf{1}_k^\top & 1 \end{bmatrix} \begin{bmatrix} \mathbf{A} & -\boldsymbol{\rho} \\ \mathbf{0}_k^\top & 1 \end{bmatrix} \begin{bmatrix} \mathbf{I}_k & \mathbf{0}_k \\ -\mathbf{1}_k^\top & 1 \end{bmatrix} = \begin{bmatrix} \mathbf{I}_k & -\boldsymbol{\rho} \\ \mathbf{0}_k^\top & 1 - \mathbf{1}_k^\top \boldsymbol{\rho} \end{bmatrix}$$

with the appropriate partitioning of the matrices, then apply the facts that the determinant of a product of matrices is the product of the determinants, and the determinant of a triangular matrix is the product of the entries along its main diagonal.

(2) Prove that, if $\mathbf{A}$ is singular, i.e. if $\det \mathbf{A} = 0 \Leftrightarrow \sum_{l=1}^k \rho_l = 1$, then, for any $w > 0$,

$$\begin{aligned} \{\boldsymbol{x} : \mathbf{A}\boldsymbol{x} > w\boldsymbol{\rho}\} \cap \mathbb{R}_{++}^k &= \left\{ \boldsymbol{x} : x_l > \rho_l \left(\sum_{m=1}^k x_m + w \right), \, l = 1, \ldots, k \right\} \cap \mathbb{R}_{++}^k \\ &\subseteq \left\{ \boldsymbol{x} : \sum_{l=1}^k x_l > \left(\sum_{l=1}^k \rho_l \right) \left(\sum_{m=1}^k x_m + w \right) \right\} \cap \mathbb{R}_{++}^k \\ &= \{\boldsymbol{x} : 0 > w\} \cap \mathbb{R}_{++}^k = \emptyset. \end{aligned}$$

(3) When $\sum_{l=1}^k \rho_l \neq 1$, $\mathbf{A}$ is non-singular, and its inverse may be obtained from the Sherman–Morrison formula to be

$$\mathbf{A}^{-1} = \mathbf{I}_k + \frac{1}{1 - \sum_{l=1}^k \rho_l} \boldsymbol{\rho}\mathbf{1}_k^\top. \tag{2.23}$$

Use this to prove that, for any $w > 0$,

$$\{\boldsymbol{x} : \mathbf{A}\boldsymbol{x} > w\boldsymbol{\rho}\} = \frac{w}{1 - \sum_{l=1}^k \rho_l} \boldsymbol{\rho} + \{\boldsymbol{v} : \mathbf{A}\boldsymbol{v} > \mathbf{0}_k\}. \tag{2.24}$$

(4) If $\sum_{l=1}^k \rho_l > 1$, prove that there is no non-zero $\boldsymbol{v} \geq \mathbf{0}_k$ such that $\mathbf{A}\boldsymbol{v} \geq \mathbf{0}_k$, because, for any non-zero $\boldsymbol{b} \geq \mathbf{0}_k$, there is no non-zero $\boldsymbol{v} \geq \mathbf{0}_k$ such that $\mathbf{A}\boldsymbol{v} = \boldsymbol{b}$. *Hint*: Show that, for any non-zero $\boldsymbol{b} \geq \mathbf{0}_k$, the vector $\boldsymbol{y} = -\mathbf{1}_k$ satisfies $\boldsymbol{y}^\top \mathbf{A} = (\sum_{l=1}^k \rho_l - 1)\mathbf{1}_k^\top \geq \mathbf{0}_k^\top$ and $\boldsymbol{y}^\top \boldsymbol{b} < 0$, and use Farkas's lemma (Schrijver, 1986, Cor. 7.1d).

(5) Apply the above result to (2.24) to prove that, if $\sum_{l=1}^k \rho_l > 1$,

$$\{\boldsymbol{x} : \mathbf{A}\boldsymbol{x} > w\boldsymbol{\rho}\} \cap \mathbb{R}_{++}^k = \emptyset.$$

Confirm that this completes the proof of Lemma 2.1.

2.5 Full solution to the canonical probability problem

We have shown how to calculate the canonical probability $\mathbb{P}\{\mathbf{A}\boldsymbol{X} > W\boldsymbol{b}\}$ when $\mathbf{A}$ is a Z-matrix and the entries of the random vector $\boldsymbol{X}$ have arbitrary independent distributions and are independent of the random variable W. However, in order to write out the full analytic expression for this canonical probability, we need the following:

(1) a simple condition to identify when the Z-matrix $\mathbf{A}$ is an M-matrix; and
(2) an analytic expression for the Laplace transform of W.

We now discuss these two issues.

2.5.1 Determining when a Z-matrix is an M-matrix

There are many different results giving conditions under which a Z-matrix can be an M-matrix. For our purposes, it is sufficient to use the following result.

LEMMA 2.7 *See Berman & Plemmons (1994, (E_{17}), Ch. 6, Thm. (2.3), p. 135). An $n \times n$ Z-matrix $\mathbf{A}$ is an M-matrix if and only if, for every $k = 1, \ldots, n$, $\det \mathbf{A}_{[k]} > 0$, where $\mathbf{A}_{[k]}$ is the $k \times k$ submatrix of $\mathbf{A}$ comprising only the entries in the first k rows and first k columns.*

Note that $\mathbf{A}_{[n]} = \mathbf{A}$. For the problems we are interested in, we shall see that the Z-matrix $\mathbf{A}$ satisfies (see, for example, Problem 2.6)

$$\det \mathbf{A}_{[1]} > \det \mathbf{A}_{[2]} > \cdots > \det \mathbf{A}_{[n]} = \det \mathbf{A}, \tag{2.25}$$

hence such a matrix $\mathbf{A}$ is an M-matrix if and only if $\det \mathbf{A} > 0$.

2.5.2 Analytic form of Laplace transform of *W*

From (2.11), W is the total interference (plus thermal noise) power at the user. Suppose we have i.i.d. Rayleigh fades on all links between the user and any BS. If there are M BSs in the entire deployment, and we condition on their distances to the user, then the Laplace transform of W may be obtained as a product of the Laplace transforms of the fades on the M links, as shown in (1.4). However, the *unconditional* Laplace transform of W, which is what we want, requires an M-dimensional integration over the joint PDF of the distances to the M BSs. Even if we were to know this joint PDF, this integral would be mathematically and computationally challenging, and certainly would not qualify as an analytic expression.

We can extricate ourselves from this situation not by focusing on simplified and unrealistic cases with small numbers of BSs, but instead by modeling, in each tier, the number of BSs as *countably infinite*, and their locations as *points of a stochastic point process*. This permits us to exploit the body of mathematical results on point processes. In particular, we shall focus on *Poisson point processes* (PPPs) and show how to derive a closed-form analytic result for the (unconditional) Laplace transform of W. Before we can do so, however, we discuss the applicability of Poisson point processes to wireless deployment scenarios, and briefly review some basic results on Poisson point processes. This forms the subject of Chapter 3.

Problems

2.6 Verify that (2.25) holds when $\mathbf{A}$ is the matrix on the LHS of (2.10). *Hint*: Write $\mathbf{A}$ in the form (2.21) and use mathematical induction.

2.7 Prove, using Lemma 2.7 and Lemma 2.1, that the probability of the event defined by (2.10) is non-zero if and only if $\sum_{l=1}^{k} \rho_l < 1$.

3 Poisson point processes

This chapter will serve as a quick review of the essential results in the theory of Poisson point processes (PPPs) that we shall use for our analysis later in the book. However, we shall first take a step back and discuss point processes in general and their applicability to our wireless deployments of interest. Then we shall provide arguments in support of the use of Poisson point processes before summarizing the most important mathematical results. This chapter is not meant to be an exhaustive treatment of the theory of PPPs – for that, see Kingman (1993). The principal theorems will be quoted with citations only to appropriate references, but heuristic arguments will be provided to help the reader understand them.

3.1 Stochastic models for BS locations

In this book, we restrict ourselves to BS deployments on a plane, i.e. in two dimensions. This models many scenarios of interest, but not all, e.g. low-power BSs ("access points") mounted indoors on several floors of an office building. Nonetheless, the theory in two dimensions is rich enough to yield useful results for a number of practical scenarios, and the techniques we employ to derive these results in this book can be extended straightforwardly to three dimensions.

A plot of the locations of the BSs in such a deployment is a so-called "spatial point pattern," or "point pattern" for short. In practice, network operators spend enormous amounts of time and resources on finding the best locations for their BS sites, taking into account such factors as the traffic they expect to support, geographical obstacles and features, municipal laws and zoning permissions, etc. However, from the point of view of a user in the network, the exact point pattern of BSs is not only unknown, but also very likely unknowable. Thus, from the point of view of this user, the locations of the BSs in the network at a given instant may be treated as *random*. Note that we do not mean that the network operator deploys the BSs randomly in the sense of throwing a dart at a map to get the BS site locations. Rather, the approach we propose may be seen as a Bayesian attempt by the user to estimate, say, the probability that the SINR at the user exceeds a fixed threshold, which requires defining a *prior* distribution on the distances to the BSs.

Proceeding with this "random" model for the locations of the BSs, it follows that the above point pattern is generated from some underlying random mechanism, i.e. a *stochastic model*. In fact, each "point" of the point pattern is an *event* of this stochastic model. The set of point patterns generated by this stochastic model is called a *point process*. In other words, a point process is identified with the set of its events.

A basic descriptor of a point process is the function that counts the number of points of the process that lie in a given region. Since the process is random, this number is a random variable. The marginal distribution of this random variable, and the joint distribution of a collection of such random variables, each denoting the number of points of the process in a different region, are important descriptors of the point process. Another useful descriptor is the distribution of the points of the process given that they all lie within some region. This lets us distinguish between model point patterns that have a "clumped" or "clustered" appearance and those that have a more "uniform" appearance. A third, and related, descriptor of the point process is the distribution of the distance from a given point of the process to the nearest other point of the process.

Note that a point process generates point patterns with certain spatial properties; each point pattern generated by the point process is called a *realization* of the point process. The study of such point processes is called *stochastic geometry*.

3.2 Complete spatial randomness

In this book, we study the simplest model for the random (as seen from the point of view of the user) locations of the BSs in the plane. In this model, the user does not know anything at all about the BS locations and makes the assumption of "maximum randomness" about their location.

(1) Given that there are $n \geq 1$ BSs in a finite region of the plane, the locations of these n BSs are i.i.d. uniformly distributed over that finite region. In other words, knowing the number of BSs in this region does not give us any information about where in the region they may be.
(2) The two random variables corresponding to the number of BSs in two disjoint regions of the plane are independent. Again, this means that knowing the number of BSs in a given region tells us nothing about the number of BSs in any other region of the plane that does not overlap the first region.

In the terminology of statistics, we seek a point process model for BS locations such that the point patterns generated by it exhibit the above properties, together called *complete spatial randomness* (CSR). It turns out that such a point process does exist (Diggle, 2003, Sec. 4.4, p. 47). Further, it is fully described by a single parameter, the constant *density* λ of points of the process, defined as the mean number of points of the process per unit area. The number of points of the process that lie in a fixed region $\mathcal{A} \subset \mathbb{R}^2$ turns out to be a *Poisson* random variable with mean $\lambda \times \text{area}(\mathcal{A})$. This point process is called the *homogeneous* (because the density is not dependent on location in the plane) *Poisson point process* (PPP). The PPP model is the most widely used model in stochastic geometry, and the basis for many classes of more sophisticated point process models.

3.3 The Poisson point process

It turns out that the discussed properties of the homogeneous PPP (i.e. with constant density λ) generalize easily to the *inhomogeneous* PPP, where the number of points of the process in a region $\mathcal{A}$ is not $\lambda \times \text{area}(\mathcal{A})$ but $\iint_{\mathcal{A}} \lambda(x,y)\mathrm{d}x\,\mathrm{d}y$, where the function $\lambda(x,y)$ of the coordinates (x,y) is now called the *intensity function* (or just *intensity*, for short) of the PPP. Note that if the PPP is homogeneous, its intensity function is constant, and the density of the PPP equals its intensity. More formally, the definition of a (possibly inhomogeneous) PPP Φ is the following (Møller & Waagepetersen, 2004, Defn. 3.2, p. 14).

(1) The number of points of the process in any finite region $\mathcal{A} \subset \mathbb{R}^2$, denoted $N(\mathcal{A})$, is a random variable with the Poisson distribution

$$\mathbb{P}\{N(\mathcal{A}) = n\} = \exp[-\mu(\mathcal{A})]\frac{[\mu(\mathcal{A})]^n}{n!}, \quad n = 0, 1, 2, \ldots,$$

with mean

$$\mu(\mathcal{A}) \equiv \iint_{\mathcal{A}} \lambda(x,y)\mathrm{d}x\,\mathrm{d}y, \quad \mathcal{A} \subseteq \mathbb{R}^2, \tag{3.1}$$

where $\lambda(\cdot,\cdot)$ is a non-negative-valued function of two variables called the *intensity function* of the PPP Φ.

(2) Given the number of points of Φ in any finite region $\mathcal{A} \subset \mathbb{R}^2$, i.e. conditioned on $N(\mathcal{A}) = n$, say, the locations of these n points are i.i.d. with PDF $\lambda(x,y)/\mu(\mathcal{A})$ over $\mathcal{A}$.

(3) For two disjoint finite regions $\mathcal{A} \subset \mathbb{R}^2$ and $\mathcal{B} \subset \mathbb{R}^2$, the corresponding numbers of points of Φ in these regions, $N(\mathcal{A})$ and $N(\mathcal{B})$, are independent.

These three properties are not independent: the third can be derived from the first two (Møller & Waagepetersen, 2004, Prop. 3.2, pp. 16–17) and the second can be derived from the first and third (Baddeley, 2007, Lem. 1.1). The function $\mu(\cdot)$ defined by (3.1) mapping subsets of the plane to $\mathbb{R}_+$ is called the *mean measure* of the PPP. It can be shown that knowledge of the mean measure of a PPP is equivalent to knowledge of its intensity function, and vice versa. Thus, either the intensity function or its integral, the mean measure, can be used to describe a PPP.

In this book, we study only BS deployments modeled by homogeneous PPPs, i.e. with a constant density on the plane. It is important to note that the PPP model is not restricted to a finite region, so in effect we are assuming that the BS deployment is infinite. This implies that a user anywhere in the network sees interference from infinitely many BSs. Of course, the slope-intercept model (1.1) ensures that the interference power decreases rapidly with distance, so each interfering BS more than a certain distance away from the user contributes very little interference. However, if there are infinitely many such BSs, one may wonder if their combined contribution can become large. As we show later in this chapter, if the BS locations are modeled by a homogeneous PPP and the path-loss exponent $\alpha > 2$, their total interference at an arbitrary location (a random variable) is not only infinite with zero probability, but also has a distribution that can be calculated analytically. Before we can prove this result, however, we need to review some basic properties and results of PPPs.

3.4 Theorems about PPPs

Two PPPs, Φ and Ψ, are said to be *independent* if, for any two finite regions $\mathcal{A}$ and $\mathcal{B}$, the random variables $N(\mathcal{A})$ and $N(\mathcal{B})$, corresponding respectively to the number of points of the PPP Φ in region $\mathcal{A}$ and the number of points of the PPP Ψ in region $\mathcal{B}$, are independent.

3.4.1 Mapping theorem

The first important result about PPPs is simple but surprisingly powerful. We state only a simplified version of the general result.

THEOREM 3.1 *(Kingman, 1993, p. 18.) Let f be a function from $\mathbb{R}^2$ to $\mathbb{R}$. Then the point process $\tilde{\Phi}$ on the real line obtained from the mapping f applied to the points of a PPP Φ with intensity function $\lambda(\cdot,\cdot)$ is a one-dimensional PPP with mean measure*

$$\tilde{\mu}(\mathcal{A}) = \iint_{\{(x,y):f(x,y)\in\mathcal{A}\}} \lambda(x,y)\mathrm{d}x\,\mathrm{d}y, \quad \mathcal{A} \subseteq \mathbb{R}, \tag{3.2}$$

provided $\tilde{\mu}(\mathcal{A})$ is finite for all bounded sets $\mathcal{A}$, and, for any $z \in \mathbb{R}$, the singleton set $\{z\}$ has zero mean measure:

$$\tilde{\mu}(\{z\}) = \iint_{\{(x,y):f(x,y)=z\}} \lambda(x,y)\mathrm{d}x\,\mathrm{d}y = 0. \tag{3.3}$$

Note that the statement of this theorem describes the resulting PPP in terms of its mean measure rather than its intensity function. Strictly speaking, the mean measure $\tilde{\mu}$ is only defined on *measurable* sets $\mathcal{A}$, i.e. sets that are countable unions and/or intersections of closed intervals on the real line. In practice, this will be true of all sets we will encounter. The condition (3.3) just states that no region with non-zero area is mapped by f onto the single point z. Clearly, the constant function $f(x,y) = c$ does not satisfy (3.3). It is also easy to see how to extend Theorem 3.1 to the case where $f: \mathbb{R}^2 \to \mathbb{R}^2$ or, in general, where $f: \mathbb{R}^d \to \mathbb{R}^s$. For brevity, we write $\tilde{\Phi} = f(\Phi)$.

Remark 3.1 An immediate consequence of Theorem 3.1 is that the projections of the coordinates of the points of the process along, say, either the x- or the y-axis also form a PPP. The same holds for the one-dimensional point process defined by the distances of points of the two-dimensional PPP from the origin (see Problem 3.1).

Remark 3.2 Another very interesting application is for a one-dimensional inhomogeneous PPP Φ with intensity function $\lambda(\cdot)$. Define the function $f: \mathbb{R} \to \mathbb{R}$ to be the area under the curve of $\lambda(\cdot)$:

$$f(x) = \int_{-\infty}^{x} \lambda(t)\mathrm{d}t, \quad x \in \mathbb{R}.$$

Then f is continuous and monotone increasing, and Theorem 3.1 shows that the mapped PPP $f(\Phi)$ is a *homogeneous* one-dimensional PPP with unit density (Kingman, 1993, p. 21). In other words, every one-dimensional PPP (homogeneous or not) can be transformed into a homogeneous one-dimensional PPP using a continuous monotone increasing transformation. Thus, when studying one-dimensional PPPs, it is sufficient in theory to focus only on homogeneous PPPs.

Consider a deployment of BSs whose locations are modeled by points of a PPP (homogeneous or inhomogeneous), all transmitting with the same power. Consider a user located at the origin and assume no fading. Then the slope-intercept path-loss model (1.1) implies that the received powers (1.2) at the user from these BSs are functions only of the distances of these BSs from the user (i.e. the origin). Since the distances of the users from the origin form a one-dimensional PPP, Theorem 3.1 says that these received powers also form a one-dimensional PPP (see Problem 3.2). This is a remarkable simplification of the original problem. We do not explore this further at this time, instead returning to it after we can account for the i.i.d. fades on the links from the BSs to the user.

Problems

3.1 Consider a PPP Φ with intensity function $\lambda(x, y)$, $(x, y) \in \mathbb{R}^2$. Let $\tilde{\Phi}$ be the one-dimensional point process whose points are the distances of the points of Φ from the origin $(0, 0)$.

(1) Show that $\tilde{\Phi} = f(\Phi)$, where $f(x, y) = \sqrt{x^2 + y^2}$, $(x, y) \in \mathbb{R}^2$.

(2) Apply Theorem 3.1 to prove that $\tilde{\Phi}$ is a PPP. *Hint*: Use the fact that, for any $r > 0$, the set $\{(x, y) \in \mathbb{R}^2 : f(x, y) = r\}$ is a circle, and the integral of $\lambda(\cdot, \cdot)$ over this circle is zero.

(3) Prove that $\tilde{\Phi}$ has mean measure given by

$$\tilde{\mu}((0, s]) = \int_0^s r \int_0^{2\pi} \lambda(r\cos\theta, r\sin\theta)\mathrm{d}\theta \, \mathrm{d}r, \quad s \geq 0.$$

Hint: Use (3.2) and transform from cartesian to polar coordinates.

(4) Prove that $\tilde{\Phi}$ is thus a one-dimensional PPP with intensity function

$$\tilde{\lambda}(r) = r \int_0^{2\pi} \lambda(r\cos\theta, r\sin\theta)\mathrm{d}\theta, \quad r \geq 0.$$

3.2 Let $\tilde{\Phi}$ be the one-dimensional PPP in Problem 3.1, and let $\Phi' = g(\tilde{\Phi})$, where $g(r) = P/r^{\alpha}$, $r \geq 0$, where $\alpha > 2$.

(1) Apply Theorem 3.1 to prove that Φ' is a PPP with mean measure given by

$$\mu'((0, y]) = \int_{(P/y)^{1/\alpha}}^{\infty} \tilde{\lambda}(r)\mathrm{d}r, \quad y \geq 0.$$

(2) Prove that the intensity function of Φ' is thus

$$\lambda'(y) = \frac{P^{2/\alpha}}{\alpha y^{1+2/\alpha}} \int_0^{2\pi} \lambda\left(\left[\frac{P}{y}\right]^{1/\alpha} \cos\theta, \left[\frac{P}{y}\right]^{1/\alpha} \sin\theta\right) \mathrm{d}\theta, \quad y \geq 0,$$

where $\lambda(\cdot, \cdot)$ is the intensity function of the PPP Φ in Problem 3.1.

(3) If Φ is a homogeneous PPP with constant intensity (density) λ, prove that Φ' is an inhomogeneous PPP with intensity function

$$\lambda'(y) = \frac{2\pi\lambda P^{2/\alpha}}{\alpha y^{1+2/\alpha}}, \quad y \geq 0. \tag{3.4}$$

3.4.2 Superposition theorem

The following theorem actually extends to a countable collection of independent PPPs, but, for our purposes, it is sufficient to state the theorem for a finite collection.

THEOREM 3.2 *(Kingman, 1993, p. 16.) Given n independent PPPs $\Phi_1, \ldots, \Phi_n$ with intensity functions $\lambda_1(\cdot,\cdot), \ldots, \lambda_n(\cdot,\cdot)$, respectively, their* superposition *Φ, defined as the point process that is the union of the sets of point patterns of the individual PPPs,*

$$\Phi = \bigcup_{i=1}^{n} \Phi_i,$$

is also a PPP and has intensity function

$$\lambda(x,y) = \sum_{i=1}^{n} \lambda_i(x,y), \quad (x,y) \in \mathbb{R}^2.$$

Note that, if the PPPs are homogeneous, this result is intuitively obvious from the assumptions of complete spatial randomness. The above theorem establishes it for inhomogeneous PPPs.

For our cellular deployment scenarios, it follows that, if the BSs in the tiers of the network (e.g. the macrocellular and picocell tiers) are modeled as points of independent PPPs, the overall point pattern generated by the superposition of all the tiers is also a PPP. In other words, the locations of all BSs in the network can be modeled by a single PPP.

3.4.3 Coloring theorem

Section 3.4.2 established that the locations of all BSs in the network can be modeled by a single PPP. Conversely, suppose we encounter, for the first time, a deployment of BSs that we are told can be modeled by a PPP. Further, we are told that each BS in this deployment belongs to one of several classes, e.g. macrocells, picocells, femtocells, etc. In other words, if we were to pick an arbitrary BS from this deployment, it may turn out to be a macrocell, say, with some probability, a picocell with some other probability, and so on. What can we say about the point process defined by all points of the overall process that belong to a certain class (e.g. macrocells)? When the overall process is a PPP, this point process is also a PPP, as the following theorem shows.

Coloring with constant retention probability

We begin with the statement of a simpler version of the general result.

THEOREM 3.3 *(Kingman, 1993, p. 53.) Let Φ be a PPP with intensity function $\lambda(\cdot,\cdot)$ and mean measure $\mu(\cdot)$. Let each point of the PPP Φ belong to exactly one of k classes (labeled $1,\ldots,k$). Let the probability that an arbitrary point of Φ is in class i be p_i, $i = 1,\ldots,k$, independent of any other point and of the locations of the points. Let Φ_i be the point process of the points of Φ belonging to class i, $i = 1,\ldots,k$. Then the point processes $\Phi_1,\ldots,\Phi_k$ are independent PPPs, with Φ_i having mean measure*

$$\mu_i(\mathcal{A}) = p_i\mu(\mathcal{A}), \quad i = 1,\ldots,k, \quad \mathcal{A} \subseteq \mathbb{R}^2, \tag{3.5}$$

or, equivalently, having intensity function

$$\lambda_i(x,y) = p_i\lambda(x,y), \quad i = 1,\ldots,k, \quad (x,y) \in \mathbb{R}^2. \tag{3.6}$$

This theorem is called the "coloring" theorem from the interpretation of the classes as being colors, so the spatial point pattern of Φ comprises points with k colors, and Φ_i is the spatial point pattern corresponding to the points of Φ with color i.

We will not give a full proof of this result, but provide the following heuristic argument: consider one of these point processes, say Φ_1. Then an arbitrary point of Φ is in Φ_1 with probability p_1 and not in Φ_1 with probability $1 - p_1$. Now consider any region $\mathcal{A}$. The number of points $N(\mathcal{A})$ of Φ in this region is a Poisson random variable with mean $\mu(\mathcal{A}) = \iint_{\mathcal{A}} \lambda(x,y)\mathrm{d}x\,\mathrm{d}y$. The number of points $N_1(\mathcal{A})$ of Φ_1 in this region is then a sum of $N(\mathcal{A})$ i.i.d. $\mathrm{Bin}(1,p_1)$ random variables that are independent of $N(\mathcal{A})$, and we can show that $N_1(\mathcal{A})$ is a Poisson random variable with mean $p_1\mu(\mathcal{A})$ (see Problem 3.3). For disjoint regions $\mathcal{A}$ and $\mathcal{B}$, the independence of $N_1(\mathcal{A})$ and $N_1(\mathcal{B})$ follows from the independence of $N(\mathcal{A})$ and $N(\mathcal{B})$. From the discussion following the definition of the PPP in Section 3.3, we know that this is sufficient to establish that Φ_1 is a PPP with intensity function $p_1\lambda(x,y)$, $(x,y) \in \mathbb{R}^2$. Regarding the independence of $\Phi_1,\ldots,\Phi_k$, suppose $k = 2$ for simplicity. It is obvious that $N_1(\mathcal{A}_1)$ and $N_2(\mathcal{A}_2)$ are independent for two disjoint regions $\mathcal{A}_1$ and $\mathcal{A}_2$, so, to prove the independence of Φ_1 and Φ_2, it suffices to prove the independence of $N_1(\mathcal{A})$ and $N_2(\mathcal{A})$ for any region $\mathcal{A}$. Note that $N(\mathcal{A}) = N_1(\mathcal{A}) + N_2(\mathcal{A})$ and $N(\mathcal{A})$ is Poisson with mean $\mu(\mathcal{A})$. The independence of $N_1(\mathcal{A})$ and $N_2(\mathcal{A})$ is proven in Problem 3.4.

The processes $\Phi_1,\ldots,\Phi_k$ in Theorem 3.3 are called *thinned* versions of Φ, and the corresponding $p_1,\ldots,p_k$ are called the *retention probabilities* of classes $1,\ldots,k$, respectively.

Coloring with a location-dependent retention probability

Note that the statement of Theorem 3.3 required the retention probability p_i for any i to be the same for every point of the PPP Φ, regardless of the location of this point. What happens if p_i depends on the location (x,y) of this point? For any $(x,y) \in \mathbb{R}^2$, define $p_i(x,y)$ to be the probability that a point of the PPP Φ located at (x,y) belongs to class i, $i = 1,\ldots,k$. A generalization of Theorem 3.3 (Møller & Waagepetersen, 2004, Prop. 3.7, p. 23) states that the point processes $\Phi_1,\ldots,\Phi_k$ are independent PPPs, with the intensity function of Φ_i given by

$$\lambda_i(x,y) = p_i(x,y)\lambda(x,y), \quad i = 1,\ldots,k, \quad (x,y) \in \mathbb{R}^2, \tag{3.7}$$

or, equivalently, the mean measure of Φ_i is

$$\mu_i(\mathcal{A}) = \iint_{\mathcal{A}} p_i(x,y)\lambda(x,y)\mathrm{d}x\,\mathrm{d}y, \quad \mathcal{A} \subseteq \mathbb{R}^2. \tag{3.8}$$

In the remainder of this book, when we mention the coloring theorem, we mean Theorem 3.3 with (3.7) replacing (3.6) and (3.8) replacing (3.5).

Remark 3.3 An interesting consequence of this result is that any inhomogeneous PPP with intensity function $\lambda(\cdot,\cdot)$ bounded by a constant c can be seen as a thinned version of a homogeneous PPP with density c using retention probability $p(x,y) = \lambda(x,y)/c$, $(x,y) \in \mathbb{R}^2$ (Møller & Waagepetersen, 2004, Cor. 3.1, p. 24).

Another useful application of Theorem 3.3 is that, instead of studying n independent PPPs Φ_i with intensity functions $\lambda_i(\cdot,\cdot)$, $i = 1,\ldots,n$, each representing one tier (class) of BSs in a multi-tier deployment, we may model the overall BS locations as points of the *single* superposed PPP Φ of Theorem 3.2 with intensity function $\lambda(x,y) = \sum_{i=1}^{n} \lambda_i(x,y)$, $(x,y) \in \mathbb{R}^2$, where any point of Φ at location $(x,y) \in \mathbb{R}^2$ belongs to tier i with probability $p_i(x,y) = \lambda_i(x,y)/\lambda(x,y)$, $i = 1,\ldots,n$, independent of all other points of Φ.

Problems

3.3 Suppose N is a Poisson random variable with mean μ, and, for any $N = n$, let $X_1,\ldots,X_n$ be i.i.d. $\mathrm{Bin}(1,p_1)$ random variables. Prove that $N_1 = \sum_{i=1}^{N} X_i$ is Poisson with mean $p_1\mu$. *Hint*: Show by conditioning on N that the moment generating function $\phi_{N_1}(s) = \mathbb{E}s^{N_1}$ is given by $\phi_{N_1}(s) = \phi_N(\phi_X(s))$, where X is $\mathrm{Bin}(1,p_1)$, then use the known forms of $\phi_X(s)$ and $\phi_N(s)$ and the fact that a distribution is uniquely identified by its moment generating function.

3.4 Define $N_2 = N - N_1$, where N and N_1 are as defined in Problem 3.3. Prove that, for any n_1 and n_2,

$$\mathbb{P}\{N_1 = n_1, N_2 = n_2\} = \mathbb{P}\{N_1 = n_1\}\,\mathbb{P}\{N_2 = n_2\},$$

where N_1 is Poisson with mean $p_1\mu$, N_2 is Poisson with mean $p_2\mu$, and $p_2 = 1 - p_1$. *Hint*: Condition on N and use the fact that, with N_1 as defined in Problem 3.3,

$$\mathbb{P}\{N_1 = n_1, N_2 = n_2 \mid N = n\} = \frac{(n_1+n_2)!}{n_1!\,n_2!} p_1^{n_1} p_2^{n_2} \delta_{n,n_1+n_2},$$

where

$$\delta_{n,m} = \begin{cases} 0, & n \neq m, \\ 1, & n = m, \end{cases}$$

is the *Kronecker delta* function.

3.4.4 Marking theorem

We begin with the observation that in the statement of the coloring theorem (Theorem 3.3 with the generalization to (3.7)), the label of the class to which any point (x, y) of the PPP Φ belongs may be viewed as a random variable $Z(x, y)$, say. Note that $Z(x, y)$ is not defined for any arbitrary location $(x, y) \in \mathbb{R}^2$, but only if the location (x, y) corresponds to a point of (the point pattern of) the PPP Φ. In (3.7), the random variable $Z(x, y)$ is discrete, taking values in $\{1, \ldots, k\}$, with *probability mass function* (PMF) conditioned on the point pattern of Φ given by

$$p_{Z(x,y)\,|\,\Phi}(i\,|\,x, y) = p_i(x, y), \quad i = 1, \ldots, k, \tag{3.9}$$

but $Z(x, y)$ may be continuous-valued, with PDF $f_{Z(x,y)\,|\,\Phi}(\cdot\,|\,x, y)$ conditioned on the point pattern of Φ. In any case, the distribution of $Z(x, y)$ given $(x, y) \in \Phi$ depends only on (x, y) and not on any other point of the PPP Φ. Further, given the point pattern of the PPP Φ, and two points (x, y) and (x', y') from this point pattern, the corresponding random variables $Z(x, y)$ and $Z(x', y')$ are independent. In the language of point processes, the random variable $Z(x, y)$ associated with each point (x, y) of the point pattern of the PPP Φ is called a *mark*, and the PPP Φ together with the set of (independent, conditioned on the point pattern) marks associated with its points is called a *marked PPP*. The *marking theorem* extends the coloring theorem to say that a marked PPP is a PPP in the product space $\mathbb{R}^2 \times \mathbb{R}$, and gives the intensity function of this PPP.

THEOREM 3.4 *(Streit, 2010, eqn. (8.2), p. 205.) With the preceding notation, the marked point process in $\mathbb{R}^2 \times \mathbb{R}$ defined by the set of points (x, y) of the PPP Φ together with their associated marks $Z(x, y)$,*

$$\tilde{\Phi} = \{(x, y, z) \in \mathbb{R}^2 \times \mathbb{R} : (x, y) \in \Phi,\ z = Z(x, y)\},$$

is a PPP on $\mathbb{R}^2 \times \mathbb{R}$ with intensity function

$$\tilde{\lambda}(x, y, z) = f_{Z(x,y)\,|\,\Phi}(z\,|\,x, y)\lambda(x, y), \quad (x, y, z) \in \mathbb{R}^2 \times \mathbb{R}, \tag{3.10}$$

or, equivalently, mean measure

$$\tilde{\mu}(\mathcal{A} \times \mathcal{B}) = \iiint_{\mathcal{A}\times\mathcal{B}} f_{Z(x,y)\,|\,\Phi}(z\,|\,x, y)\lambda(x, y)\mathrm{d}x\,\mathrm{d}y\,\mathrm{d}z, \quad \mathcal{A} \subseteq \mathbb{R}^2,\ \mathcal{B} \subseteq \mathbb{R}. \tag{3.11}$$

Remark 3.4 In the statement of Theorem 3.4, we took care to write the space of the marked PPP $\tilde{\Phi}$ as the product space $\mathbb{R}^2 \times \mathbb{R}$ instead of just $\mathbb{R}^3$ because, in general, the marks Z could be random vectors instead of scalar random variables, and the statement of the theorem readily generalizes to this case. Though we use the notation for the PDF $f_{Z(x,y)\,|\,\Phi}(\cdot\,|\,x, y)$ in (3.10), we do not restrict the statement of Theorem 3.4 to apply only to continuous-valued Z. In fact, the coloring theorem (Theorem 3.3) is a special case of the marking theorem (Theorem 3.4), as is shown in Problem 3.5.

Problem

3.5 Derive (3.8) as a special case of (3.11). *Hint*: First show that $\mu_i(\mathcal{A})$ in (3.8) is just $\tilde{\mu}(\mathcal{A} \times \{i\})$ in (3.11), then evaluate $\tilde{\mu}(\mathcal{A} \times \{i\})$ using (3.9) and $f_{Z(x,y)\,|\,\Phi}(z\,|\,x, y) =$

$\sum_{i=1}^{k} p_{Z(x,y)\,|\,\Phi}(i\,|\,x,y)\delta(z-i)$, where $\delta(\cdot)$ is the *Dirac delta* function, i.e. the derivative of the *unit step* function

$$U(x) = \begin{cases} 1, & x \geq 0, \\ 0, & x < 0. \end{cases}$$

3.5 Applicability of PPP to real-world deployments

We shall develop tools for analyzing the distribution of the SINR in multi-tier HCNs where the locations of the BSs in the tiers are modeled as points of independent homogeneous PPPs. However, as we have seen, the essential feature of the PPP is its modeling of complete spatial randomness, and this is exactly opposed to the goals of network design. In particular, a network operator most definitely does not want to sprinkle BSs over the deployment region independently and at random.

Consider the deployment of a single-tier network. If traffic is uniform across the deployment region, the classic "hexagonal grid" layout of BSs is a good solution, as it reduces the deployment problem to the choice of a single parameter, the so-called *inter-site distance* (ISD), i.e. the distance between nearest-neighbor BSs. However, traffic is seldom, if ever, uniform over a deployment region of any significant size, which means that there may be more BSs, with smaller range, in certain regions, and fewer BSs, with larger range, in other regions of the deployment area. Ideally, for a so-called greenfield deployment (i.e. there was no network there before, hence a "green field"), the operator estimates traffic demand over the region, then tries to optimize the placement of BSs so as to achieve an optimal tradeoff between cost, coverage, capacity, and traffic load distribution across the network. As if this optimization is not hard enough already, it is still an idealized problem, because it assumes that the BSs can be placed anywhere in the region. This is, of course, far from true, with numerous access, zoning, and right-of-way restrictions bedeveling the operator's choices of BS sites.

In other words, various constraints eventually drive the network operator toward a BS placement over a given deployment area that is often very different from that of the classic "hexagonal grid" layout. Is a PPP spatial point pattern a better model for the kinds of real-world deployments that do occur?

Consider a user located at the origin of an arbitrary single-tier deployment with i.i.d. log-normal fading on all links between BSs of this deployment and this user. Let H be a random variable with the same distribution as the common distribution of the fade attenuations on these links, i.e.

$$H = \exp(\sigma Z - \sigma^2/2), \quad Z \sim \mathcal{N}(0, 1).$$

A recent result (Błaszczyszyn, Karray & Keeler, 2013, Thm. 3) shows that if the (deterministic) network of BSs has *empirical homogeneity* (meaning that the empirical density of BSs over a disk of radius r centered at the user tends to a finite positive value λ, say, as $r \to \infty$), then the one-dimensional point process of received powers at the user from the BSs in this deployment converges in distribution to the one-dimensional PPP of received powers at the user from the BSs of a homogeneous PPP deployment with density λ and intensity function (4.6) with no fading as $\sigma \to \infty$. It follows that (Błaszczyszyn *et al.*, 2013, Thm. 3, Rem. 4) a homogeneous PPP model for the locations of the BSs in this single tier is sufficient to approximate the distribution of the SINR at an arbitrarily located user in the real-world deterministic deployment, at least when there is only i.i.d. log-normal fading on all links to the user, and the standard deviation of the log-normal fading is sufficiently large.

Further (Błaszczyszyn *et al.*, 2013, Thm. 3), the homogeneous PPP model is also sufficient to approximate the distribution of the SINR at the arbitrarily located user even if we only consider the one-dimensional point process of received powers at the user from the BSs of the original deterministic deployment that are at some minimum distance from the user and/or less than some maximum distance from the user. In other words, the applicability of the homogeneous PPP model extends to deterministic but finite layouts. It follows that the simple path loss model (1.1), which yields infinite received power at zero distance from a BS, can be used without modification with the PPP model for BS locations, since the results obtained from such a model are applicable even if we require a minimum distance to the nearest BS in the real-world deployment (Błaszczyszyn *et al.*, 2013, Rem. 5).

This result, though powerful and general, has been shown for only one kind of fade distribution, namely for log-normal fading, and only asymptotically. Unfortunately, no similar analytic results exist for other fading models in deterministic deployments. However, it has been seen via simulation (Andrews *et al.*, 2011, Fig. 6) that, for a real-world deployment of BSs with assumed i.i.d. log-normal fade attenuation H on all links to the desired user, even for σ as low as 6 dB, an excellent approximation to the SINR distribution at an arbitrary user is obtained from a PPP BS location model with i.i.d. log-normal fade attenuation with the same distribution as H for all non-serving BSs and Rayleigh fading on the link to the user from the serving (nearest) BS.

For i.i.d. Rayleigh fading on all links in single-tier (Andrews *et al.*, 2011, Fig. 4) and two-tier (Dhillon *et al.*, 2012, Figs. 7–8) networks, simulation results show that, for a particular real-world HCN deployment of BSs in a wide area, the CCDF of maximum SINR (across all BSs in the deployment) as a function of the SINR threshold is more or less consistently underestimated as much by the PPP model as it is overestimated by a "square grid" (Andrews *et al.*, 2011) or "hexagonal grid" (Dhillon *et al.*, 2012) layout model, respectively, for each value of the threshold SINR between -10 dB and 20 dB. It is not surprising that the PPP results are conservative, i.e. they underestimate the actual CCDF (the probability that the SINR exceeds the threshold), because the PPP model imposes no restrictions on how closely two points of the process can be located. This leads to high values of interference from non-serving BSs (Mukherjee, 2013). On the other hand, it is also to be expected that the "square grid" and the "hexagonal grid" layout will be too optimistic and overestimate the actual probability that the SINR exceeds the threshold, because these models impose a minimum distance between nearest-neighbor BSs.

At any rate, one may conclude that the PPP model yields results that are, if anything, conservative. If used at the preliminary stage of network layout in order to obtain approximate values of deployment operating parameters, the PPP model should therefore yield a robust system design.

In multi-tier deployments with small cells, as the HCNs of the future are likely to be, the PPP model may be even more appropriate, because the number of cells is too large for the operator to lay out the BS sites using the traditional tools for wide-area macrocellular single-tier deployments. Moreover, the new generations of small cells tend to be relatively inexpensive and much less visually obtrusive than large macro-BS sites, and therefore can be, and usually are, placed anywhere that permission may be obtained (often from private landlords, owners of buildings on whose walls and roofs these small BSs are attached) to place them. It would seem, therefore, that the PPP location model is probably more accurate for small cell tiers than for macrocellular tiers. However, there are no studies at the present time to confirm or refute this hypothesis.

3.6 Other models for BS locations

The homogeneous PPP is the simplest stochastic model for a spatial point pattern. Numerous other models, usually trading off mathematical tractability for a closer approximation of some aspect of a real-world deployment, have been studied by researchers in the field of stochastic geometry. The most common perceived failing of the PPP is that it imposes no minimum restriction on the distance between points of the process, and stochastic models such as the Matérn hard-core process (MHP) have been proposed to remedy this. We shall not study any processes other than the PPP in this book. An overview of point processes for modeling of wireless networks may be found in Andrews *et al.* (2010). The reader is also referred to Haenggi (2012) for a detailed exposition not only of the PPP, but also a variety of other point processes, including the MHP.

4 SINR analysis for a single tier with fixed power

4.1 Introduction

We are interested in the distribution of *instantaneous* SINR at an arbitrary user location in a multi-tier deployment of BSs, where the locations of the BSs of each tier are modeled as points of a PPP. In this chapter, we restrict ourselves to a simpler problem, namely a single-tier BS deployment. We show how to derive the distribution of the SINR at the user location (assumed, without loss of generality, to be the origin) under different serving BS selection criteria (also called *association rules*, because they determine the BSs to which the user associates itself). The derivations will use results from the theory of Poisson point processes in Chapter 3. We shall also require some advanced results from the theory of PPPs that were not covered in Chapter 3, and these results will be introduced as needed in this chapter.

Before we study the distribution of instantaneous SINR at a user, let us begin with an even simpler problem, namely the distribution of the instantaneous total received power at the user from all BSs in the tier.

4.2 Distribution of total interference power in a single-tier BS deployment

4.2.1 PPP of received powers at user from BSs in a tier

Consider a marked PPP Φ with intensity function $\lambda(\cdot,\cdot)$, where the marks associated with the points (x,y) of Φ are i.i.d. with common PDF $f_H(\cdot)$, say. For brevity, let us denote an arbitrary point of Φ by b (instead of (x,y) as above), and let the mark associated with $b \in \Phi$ be denoted H_b (instead of $Z(x,y)$ as in Section 3.4.4). Suppose the points of Φ are (the locations of) the BSs in a deployment, each transmitting with power P^{tx}. Consider a user at the origin $(0,0)$, and denote the distance of a BS $b \in \Phi$ from the origin by $R_b = \|b\|$. From Theorem 3.4, we know that the marked PPP $\{(b,H_b) : b \in \Phi, H_b \text{ i.i.d.} f_H(\cdot)\}$ is a PPP on $\mathbb{R}^2 \times \mathbb{R}_+$. Next, define the function $f : \mathbb{R}^2 \times \mathbb{R}_+ \to \mathbb{R}_+$ to be the mapping from each point (b,H_b) of the marked PPP to the instantaneous received power at a user at the origin from a BS at b when the fade attenuation on the link is H_b:

$$f(b,H_b) = \frac{PH_b}{R_b^{\alpha}}, \quad P = KP^{\text{tx}}, \quad b \in \Phi, \quad \{H_b\}_{b\in\Phi} \text{ i.i.d.} f_H(\cdot), \tag{4.1}$$

where α and K are, respectively, the slope and intercept of the path-loss model (1.1), and assumed the same on all links from BSs in the deployment to the user at the origin. Then, from Theorem 3.1, it follows that the one-dimensional point process of instantaneous received powers at the user from the BSs in the deployment,

$$\tilde{\Phi} = \{y \in \mathbb{R}_+ : y = f(b, H_b), b \in \Phi, H_b \text{ i.i.d. } f_H(\cdot)\}, \tag{4.2}$$

is a one-dimensional PPP with mean measure (see Problem 4.1)

$$\tilde{\mu}((0, y]) = \int_0^\infty r \int_0^{2\pi} \lambda(r\cos\theta, r\sin\theta) \mathrm{d}\theta \, \mathrm{d}r \int_0^{yr^\alpha/P} f_H(t)\mathrm{d}t, \quad y \geq 0, \tag{4.3}$$

or, equivalently, intensity function

$$\tilde{\lambda}(y) = \frac{\mathrm{d}}{\mathrm{d}y}\tilde{\mu}((0, y]) = \int_0^\infty \frac{r^{1+\alpha}}{P} f_H\left(\frac{yr^\alpha}{P}\right) \int_0^{2\pi} \lambda(r\cos\theta, r\sin\theta)\mathrm{d}\theta \, \mathrm{d}r, \quad y \geq 0.$$

Further, if Φ is homogeneous with constant intensity (density) λ, we have

$$\tilde{\lambda}(y) = 2\pi\lambda \int_0^\infty \frac{r^{1+\alpha}}{P} f_H\left(\frac{yr^\alpha}{P}\right)\mathrm{d}r \tag{4.4}$$

$$= \frac{2\pi\lambda P^{2/\alpha}}{\alpha y^{1+2/\alpha}} \int_0^\infty t^{2/\alpha} f_H(t)\mathrm{d}t \tag{4.5}$$

$$= \frac{2\pi(\lambda\, \mathbb{E}[H^{2/\alpha}])P^{2/\alpha}}{\alpha y^{1+2/\alpha}}, \quad y \geq 0, \tag{4.6}$$

where, in (4.5), we make the change of variables $t = yr^\alpha/P$, and in (4.6) H is a random variable with the PDF $f_H(\cdot)$. Comparing (4.6) with (3.4), we obtain the following remarkable result.

THEOREM 4.1 *For the path-loss model* (1.1) *and i.i.d.* arbitrary fading *with PDF* $f_H(\cdot)$ *on all links from BSs to the user, the one-dimensional PPP of instantaneous received powers at the user location (origin) from the BSs located at points of a homogeneous PPP* Φ *with density* λ*, all transmitting with power* P^{tx}*, is* equivalent *to the one-dimensional PPP of instantaneous received powers* in the absence of fading *from the BSs located at points of a homogeneous PPP with density* $\lambda\,\mathbb{E}[H^{2/\alpha}]$*, also transmitting with the same power* P^{tx}.

Remark 4.1 Theorem 4.1 was proved for the case of i.i.d. Rayleigh fading in Haenggi (2008, Prop. 3), and in the general form stated above in Madhusudhanan *et al.* (2009, Thm. 2), and Błaszczyszyn *et al.* (2010, Prop. 5.5).

Remark 4.2 The result in Theorem 4.1 has been extended to the case where the path-loss model replaces "distance in meters" in (1.1) with max{reference distance, distance in meters} (Vu, Decreusefond & Martins, 2012). It has also been extended to the cases where the BSs in Φ: (i) transmit with i.i.d. random powers (Madhusudhanan *et al.*, 2011, Rem. 5), or (ii) have ideal sectorized antennas with i.i.d. orientation (Madhusudhanan *et al.*, 2011, Cor. 5). However, these results may not be as useful as Theorem 4.1 because the assumption of independence of either transmit powers or beam orientation across the BSs is unrealistic in a cellular network.

Remark 4.3 It is important to note that, in Theorem 4.1, the distribution of H is allowed to be arbitrary, provided $\mathbb{E}[H^{2/\alpha}] < \infty$. Further, from (4.6) we see that the distribution of H influences the PPP of received powers *only* through $\mathbb{E}[H^{2/\alpha}]$.

4.2.2 Distribution of total received power from all BSs in a tier

Let us consider a deployment of BSs whose locations are modeled as points of a homogeneous PPP Φ with density λ. Suppose all BSs transmit with fixed transmit power P^{tx} and the attenuations due to fading on all links from the BSs to the user location at the origin are i.i.d. with common PDF $f_H(\cdot)$.

Distribution of instantaneous total received power

From Theorem 4.1, we know that the instantaneous received powers at the user location from the BSs in Φ form a one-dimensional inhomogeneous PPP $\tilde{\Phi}$ with intensity function given by (4.6). The instantaneous total received power at the user from all the BSs in Φ is thus given by

$$W = \sum_{y \in \tilde{\Phi}} y. \tag{4.7}$$

The Laplace transform of W, given by $\mathbb{E}e^{-sW}$, can be calculated using the following result.

LEMMA 4.2 *(Kingman, 1993, eqn. (3.29), p. 31.) If Ψ is a PPP on $\mathbb{R}^d$ with intensity function $\nu(\cdot)$ and $f : \mathbb{R}^d \to \mathbb{R}_+$, then, for any $s > 0$,*

$$\mathbb{E} \exp\left(-s \sum_{x \in \Psi} f(x)\right) = \exp\left\{-\int_{\mathbb{R}^d} \left[1 - e^{-sf(x')}\right] \nu(x') dx'\right\}. \tag{4.8}$$

The quantity on the left of (4.8) is called the *characteristic functional* of the PPP, evaluated at the function f (recall that a *functional* is a scalar-valued function taking a function as its argument). It is a special case of the so-called *probability generating functional* (or just *generating functional*, for short)

$$\mathbb{E}\left[\prod_{x \in \Psi} g(x)\right], \quad g : \mathbb{R}^d \to (0, 1], \tag{4.9}$$

with $g(x) = \exp[-sf(x)]$, and (4.8) is a special case of the result (Kingman, 1993, eqn. (3.35), p. 33)

$$\mathbb{E}\left[\prod_{x \in \Psi} g(x)\right] = \exp\left\{-\int_{\mathbb{R}^d} [1 - g(x')] \nu(x') dx'\right\}, \tag{4.10}$$

which holds for any $g : \mathbb{R}^d \to \mathbb{R}$ provided the integral on the right of (4.10) is finite. A heuristic argument for (4.10) is provided in Problem 4.2.

Returning to the calculation of the Laplace transform of W as defined in (4.7), we can apply (4.8), replacing Ψ by $\tilde{\Phi}$, ν by $\tilde{\lambda}$ (given by (4.6)), d by 1, and f by the identity, to obtain

$$\begin{aligned}\mathbb{E}e^{-sW} &= \exp\left\{-\int_0^\infty [1-e^{-sx}]\tilde{\lambda}(x)dx\right\}\\ &= \exp\left\{-\frac{2\pi\lambda P^{2/\alpha}\,\mathbb{E}[H^{2/\alpha}]}{\alpha}\int_0^\infty \frac{1-e^{-sx}}{x^{1+2/\alpha}}dx\right\}\\ &= \exp\left\{-\pi\lambda P^{2/\alpha}\,\mathbb{E}[H^{2/\alpha}]\Gamma\left(1-\frac{2}{\alpha}\right)s^{2/\alpha}\right\}, \quad s>0,\ \alpha>2, \end{aligned} \tag{4.11}$$

where the steps in the integration are outlined in Problem 4.3, and

$$\Gamma(x) = \int_0^\infty t^{x-1}\,e^{-t}\,dt \tag{4.12}$$

is the *gamma* function. The Laplace transform expression (4.11) was first derived for the case of i.i.d. Rayleigh fading in Hunter, Andrews & Weber (2008, eqn. (6)), with an extension to Nakagami fading.

In Jacquet (2008, Thm. 2), the Laplace transform (4.11) is inverted to yield the CCDF of W, the instantaneous total received power from all BSs in the deployment:

$$\begin{aligned}&\mathbb{P}\{W>w\}\\ &= \sum_{k=1}^{\infty}(-1)^{k-1}\left[\pi\lambda P^{2/\alpha}\,\mathbb{E}[H^{2/\alpha}]\Gamma\left(1-\frac{2}{\alpha}\right)\right]^k \frac{\sin(2\pi k/\alpha)}{\pi}\frac{\Gamma(2k/\alpha)}{k!}w^{-2k/\alpha},\\ &w>0.\end{aligned}$$

Remark 4.4 By formally setting the argument to be imaginary ($s=-j\omega$) in (4.11), we obtain the characteristic function of W, whose form shows that W has an *α-stable* distribution (Ito, 1993, Sec. 5.F, p. 18) with *characteristic exponent* $2/\alpha$ and *dispersion* $\pi\lambda P^{2/\alpha}\,\mathbb{E}[H^{2/\alpha}]\Gamma(1-2/\alpha)$. This result was first derived for i.i.d. Rayleigh fading in Pinto & Win (2010a, eqn. (15)), then extended to arbitrary fading in Gulati *et al.* (2010, eqn. (15)).

Remark 4.5 The approach of starting with the Laplace transform of W (which can be inverted analytically) instead of the characteristic function of W (which cannot be inverted analytically) is the key to obtaining the CCDF of W for arbitrary fading. This may be seen by contrasting the approach of Jacquet (2008) with that of Dahama, Sowerby & Rowe (2009), where the characteristic function is obtained for specific kinds of fading (Rayleigh, log-normal, and Suzuki) and then numerically inverted. This "characteristic function followed by numerical inversion" approach is also employed in Ge *et al.* (2011, eqns. (3a), (3d), (5), (7a)), where the above result was extended from omnidirectional single antennas at the BSs and the user to the case where there are multiple antennas at both the BSs and the user, with multiple-input multiple-output (MIMO) transmission.

Remark 4.6 An alternative expression (4.15) for $\mathbb{E}e^{-sW}$ is derived in Problem 4.4. Although not as explicit as (4.11), it is useful for proving other results (see Remark 4.7 and Remark 4.8).

Distribution of total smoothed received power

The expression (4.11) for the Laplace transform of the instantaneous total received power at the user from the BSs holds for any distribution of the arbitrary fade attenuations on the links from the BSs to the user. For a specific scenario, it is possible to apply this result to derive the Laplace transform of the total received power at the user obtained after smoothing several successive instantaneous measurements, as shown in this section.[1] This is the only result in this book where we set aside our underlying assumption that we are given only an *instantaneous snapshot* of the deployment. Suppose there is i.i.d. Rayleigh fading on all links from all BSs to the user, and suppose also that the BSs transmit i.i.d. symbols from an M-ary PSK (phase shift keying) constellation that is independent of the fading. Then the in-phase and quadrature components of the instantaneous signal received at the user from each BS are independent and Gaussian with zero mean. Conditioned on the distances of all the BSs from the user location, the in-phase and quadrature components of the total instantaneous signal received at the user are also independent and zero-mean Gaussian (Pinto & Win, 2010b, eqn. (3)). Suppose we can ignore thermal noise. The instantaneous received power at the user is the sum of the squares of the in-phase and quadrature components of the total instantaneous signal received at the user. Suppose now that the user performs a smoothing of successive instantaneous total received power measurements, to obtain the smoothed total received power measurement at (discrete) time t given by

$$\sum_{\tau=t-M}^{t} h_{t-\tau} \left[X_{\text{tot,I}}^2(\tau) + X_{\text{tot,Q}}^2(\tau) \right],$$

where $h_0, \ldots, h_M$ are the coefficients (assumed real) of the filter, M is the filter window, and $X_{\text{tot,I}}(t)$ and $X_{\text{tot,Q}}(t)$ are, respectively, the in-phase and quadrature components of the total instantaneous received signal at the user at time t. Now let $X_{b,\text{I}}(t)$ and $X_{b,\text{Q}}(t)$, respectively, denote the in-phase and quadrature coefficients of the instantaneous fading on the link to the user from the BS located at b, at time t. Assume the user is stationary, so its distance from any BS does not change with time. Then we have

$$X_{\text{tot,I}}(t) = \sum_{b \in \Phi} \frac{\sqrt{P}}{R_b^{\alpha/2}} X_{b,\text{I}}(t), \quad X_{\text{tot,Q}}(t) = \sum_{b \in \Phi} \frac{\sqrt{P}}{R_b^{\alpha/2}} X_{b,\text{Q}}(t),$$

where we recall that, at any t and for any $b \in \Phi$, $X_{b,\text{I}}(t)$ and $X_{b,\text{Q}}(t)$ are i.i.d. $\mathcal{N}(0, \frac{1}{2})$. Suppose also that the smoothing filter $[h_0, \ldots, h_M]$ is such that, for any b and any time t,

$$\sum_{\tau=t-M}^{t} h_{t-\tau} X_{b,\text{I}}(\tau) \approx \mathbb{E}[X_{b,\text{I}}(t)] = 0, \quad \sum_{\tau=t-M}^{t} h_{t-\tau} X_{b,\text{Q}}(\tau) \approx \mathbb{E}[X_{b,\text{Q}}(t)] = 0.$$

[1] The remainder of this section may be skipped on a first reading.

Then we see that the smoothed total received power measurement at time t is given by

$$\begin{aligned} W_M(t) &= \sum_{\tau=t-M}^{t} h_{t-\tau}\left[X_{\mathrm{tot,I}}^2(\tau) + X_{\mathrm{tot,Q}}^2(\tau)\right] \\ &\approx \sum_{b\in\Phi} \frac{P}{R_b^\alpha}\left\{\sum_{\tau=t-M}^{t} h_{t-\tau}\left[X_{b,\mathrm{I}}^2(\tau) + X_{b,\mathrm{Q}}^2(\tau)\right]\right\} \\ &= \sum_{b\in\Phi} \frac{P}{R_b^\alpha}\left(\sum_{\tau=t-M}^{t} h_{t-\tau} H_b(\tau)\right), \end{aligned} \tag{4.13}$$

where in the final step we defined the fade power attenuation at time t by $H_b(t) = X_{b,\mathrm{I}}^2(t) + X_{b,\mathrm{Q}}^2(t)$. Comparing (4.13) with (4.7), we see that the Laplace transform of the smoothed total received power $W_M(t)$ at any time t is given by the expression (4.11) for the Laplace transform of W as defined by (4.7), with the random variable H defined by $H = \sum_{\tau=t-M}^{t} h_{t-\tau} H_b(\tau)$. The difficulty is in calculating the distribution of this H, which we discuss next.

Recall that our assumption of i.i.d. Rayleigh fading means that $H_b(t) \sim \mathrm{Exp}(1)$ for each t, and for any t, t' and $b \neq b'$, $\mathbb{E}[H_b(t)H_{b'}(t')] = \mathbb{E}[H_b(t)]\,\mathbb{E}[H_{b'}(t')] = 1$, but we have not said anything so far about the autocorrelation $\mathbb{E}[H_b(t)H_b(t')]$ for a given b when $t \neq t'$. It turns out that, for a completely general autocorrelation function, the *Laplace transform* $\mathcal{L}_H(\cdot)$ of the random variable H as defined previously can be obtained in closed form for several different choices of smoothing filter coefficients $h_0, \ldots, h_M$ (e.g. Kleptsyna, Le Breton & Viot (2002, eqn. (28))). While this does not give us $\mathbb{E}[H^{2/\alpha}]$ to use in (4.11), the Laplace transform $\mathcal{L}_H(\cdot)$ can be used in (4.15).

Problems

4.1 Prove that (4.2) is a PPP with mean measure (4.3). *Hint*: First apply Theorem 3.4 and (3.10) to the marked PPP $\{(b, H_b) : b \in \Phi, H_b \text{ i.i.d. } f_H(\cdot)\}$ to show that its intensity function at $(x, y, h) \in \mathbb{R}^2 \times \mathbb{R}_+$ is $f_H(h)\lambda(x, y)$. Next, switch to polar coordinates and prove that, for any $y \geq 0$, the set $\{(r\cos\theta, r\sin\theta, h) : r \in \mathbb{R}_+, \theta \in [0, 2\pi), h \in \mathbb{R}_+, Ph/r^\alpha = y\}$ has no volume, then apply (3.2) to the marked PPP.

4.2 The *void probability* over a region $\mathcal{A} \subseteq \mathbb{R}^d$ of a PPP Ψ on $\mathbb{R}^d$ with intensity function $\nu(\cdot)$ is the probability that there is no point of Ψ in the region $\mathcal{A}$.

(1) Prove that the void probability of Ψ over $\mathcal{A} \subseteq \mathbb{R}^d$ is $\exp[-\mu(\mathcal{A})]$, where $\mu(\mathcal{A}) = \int_{\mathcal{A}} \nu(\boldsymbol{x})\mathrm{d}\boldsymbol{x}$ is the mean measure of Ψ.

(2) Given a function $g : \mathbb{R}^2 \to (0, 1]$, consider the PPP $\tilde{\Psi}$ obtained by thinning Ψ with retention probability $p(\boldsymbol{x}') = 1 - g(\boldsymbol{x}')$ at every $\boldsymbol{x}' \in \mathbb{R}^d$. Prove that the probability that $\tilde{\Psi}$ is empty (void) is

$$\exp\left\{-\int_{\mathbb{R}^d} [1 - g(\boldsymbol{x}')]\nu(\boldsymbol{x}')\mathrm{d}\boldsymbol{x}'\right\}.$$

Hint: Use (3.8) with $\mathcal{A} = \mathbb{R}^d$.

(3) Prove (4.10) by directly calculating the probability that $\tilde{\Psi}$ is void as the left side of (4.10). *Hint*: This probability is just the probability that no point of Ψ is in $\tilde{\Psi}$,

and, for each point $\boldsymbol{x} \in \Psi$, the probability that it is not in $\tilde{\Psi}$ is $g(\boldsymbol{x})$, independent of all other points of Ψ. So the desired probability is just $\mathbb{E}\left[\prod_{\boldsymbol{x}\in\Psi} g(\boldsymbol{x})\right]$.

4.3 Prove (4.11) as follows.

(1) Show that

$$\frac{2}{\alpha}\int_0^\infty \frac{1-\mathrm{e}^{-sx}}{x^{1+2/\alpha}}\,\mathrm{d}x = -\left.\frac{1-\mathrm{e}^{-sx}}{x^{2/\alpha}}\right|_0^\infty + s\int_0^\infty x^{-2/\alpha}\,\mathrm{e}^{-sx}\,\mathrm{d}x.$$

Hint: Use integration by parts.

(2) Prove that

$$\lim_{x\to 0+}\frac{1-\mathrm{e}^{-sx}}{x^{2/\alpha}} = 0, \quad \alpha > 2. \tag{4.14}$$

Hint: Use l'Hôpital's rule.

(3) Use (4.12) to complete the derivation of (4.11).

4.4 For W as defined in (4.7), with $\{H_b\}_{b\in\Phi}$ i.i.d. $f_H(\cdot)$, prove that (Hunter *et al.*, 2008, eqn. (5))

$$\mathcal{L}_W(s) = \mathbb{E}\,\mathrm{e}^{-sW} = \exp\left\{-2\pi\lambda\int_0^\infty r\left[1-\mathcal{L}_H\left(\frac{Ps}{r^\alpha}\right)\right]\mathrm{d}r\right\}, \quad s > 0. \tag{4.15}$$

Hint: Proceed as follows.

(1) As $\{H_b\}_{b\in\Phi}$ are i.i.d., show by conditioning on Φ that

$$\begin{aligned}
\mathbb{E}_{\Phi,\{H_b\}_{b\in\Phi}}\exp\left(-s\sum_{b\in\Phi}\frac{PH_b}{R_b^\alpha}\right) &= \mathbb{E}_{\Phi,\{H_b\}_{b\in\Phi}}\left[\prod_{b\in\Phi}\exp\left(-s\sum_{b\in\Phi}\frac{PH_b}{R_b^\alpha}\right)\right] \\
&= \mathbb{E}_\Phi\left\{\mathbb{E}_{\{H_b\}_{b\in\Phi}}\left[\prod_{b\in\Phi}\exp\left(-s\sum_{b\in\Phi}\frac{PH_b}{R_b^\alpha}\right)\mid\Phi\right]\right\} \\
&= \mathbb{E}_\Phi\left[\prod_{b\in\Phi}\mathcal{L}_{H_b}\left(\frac{sP}{R_b^\alpha}\right)\right] = \mathbb{E}_\Phi\left[\prod_{b\in\Phi}\mathcal{L}_H\left(\frac{sP}{R_b^\alpha}\right)\right],
\end{aligned} \tag{4.16}$$

where H is a random variable with the same distribution as $\{H_b\}_{b\in\Phi}$.

(2) Evaluate (4.16) by applying (4.10) to the PPP $\Psi = \{R_b : b \in \Phi\}$ with $g(r) = \mathcal{L}_H(sP/r^\alpha)$, $r \geq 0$. Recall from Problem 3.1 that Ψ has intensity function $\nu(r) = 2\pi\lambda r$, $r \geq 0$.

4.3 Distribution of SINR in a single-tier BS deployment

In this section, we apply the results on PPPs from Chapter 3 to derive the distribution of instantaneous SINR at an arbitrary user (assumed located at the origin) in a single-tier deployment of BSs (assumed to be transmitting with fixed power) under several scenarios. The results on SINR distribution derived here will serve as a preview of the more general results for multi-tier heterogeneous networks in the subsequent chapters. The first scenario we analyze here is when the serving BS for the user is fixed.

4.3.1 Serving BS known and fixed

Consider again a single tier of BSs whose locations are modeled as the points of a homogeneous PPP Φ with density λ. From (2.5), we see that we have the CCDF of the SINR at a given user location (the origin, say) conditioned on the distance to the serving BS provided we can compute the Laplace transform of V as defined in (2.2) for tier $i = 1$. From (2.2) and (4.7), V is the total received power W from all BSs in the deployment minus the received power U from the serving BS. Suppose the serving BS is at location (x_0, y_0), so that its distance from the user (assumed to be located at the origin) is $R_* = \sqrt{x_0^2 + y_0^2}$. It turns out that, conditioned on $(x_0, y_0) \in \Phi$, the Laplace transform of V is the *same* as the Laplace transform (4.11) of W because of *Slivnyak's theorem*, which we state informally in Theorem 4.3 (a rigorous statement relies on the definition of so-called *Palm probabilities*, which we do not go into in this book – see Baccelli & Błaszczyszyn (2009, Thm. 1.13, p. 30) for details).

Application of Slivnyak's theorem

THEOREM 4.3 *Consider any (possibly inhomogeneous) PPP Φ on $\mathbb{R}^2$ with mean measure μ. Conditioned on there being a point of Φ at the location (x_0, y_0), the point process with point pattern $\Phi \setminus \{(x_0, y_0)\}$ is a PPP on $\mathbb{R}^2 \setminus \{(x_0, y_0)\}$ with mean measure μ.*

Recall that a point process is identified with its spatial point pattern. Intuitively, it is easy to see why Theorem 4.3 holds, because the event that there is a point of Φ at (x_0, y_0) may be viewed as the event that there is exactly one point of Φ in a disk centered at (x_0, y_0) with vanishingly small radius. Then, from the definition of the PPP, the point pattern of the points of Φ that are outside this disk is independent of the point of Φ inside the disk, so the mean measure of the PPP defined by the point pattern $\Phi \setminus \{(x_0, y_0)\}$ conditioned on the point of Φ at (x_0, y_0) is the same as the unconditional mean measure of Φ.

For our purposes, it is enough to interpret Theorem 4.3 as saying that, conditioned on there being a point of a PPP Φ at a given location (x_0, y_0), the point process $\Phi' = \Phi \setminus \{(x_0, y_0)\}$ defined by the point pattern comprising the other points of the PPP is *equivalent* to the original PPP (Baccelli & Błaszczyszyn, 2009, Cor. 1.14, p. 32). It follows that the corresponding marked PPPs $\{(b, H_b) : b \in \Phi', H_b \text{ i.i.d. } f_H(\cdot)\}$ and $\{(b, H_b) : b \in \Phi, H_b \text{ i.i.d. } f_H(\cdot)\}$ are also equivalent. Applying Theorem 3.1 with the mapping f defined by (4.1) shows that the PPP $\tilde{\Phi}' = \{y \in \mathbb{R}_+ : y = f(b, H_b), b \in \Phi', H_b \text{ i.i.d. } f_H(\cdot)\}$ is equivalent to the PPP $\tilde{\Phi}$ defined in (4.2). Then the quantity V in (2.2), which is equivalent to $\sum_{y \in \tilde{\Phi}'} y$, has the same distribution as W in (4.7). From (4.11), we then have the following result.

THEOREM 4.4 *Suppose that the fade attenuations on the links to the user from all BSs in the deployment are i.i.d. with common PDF $f_H(\cdot)$. Given that the user is served by the*

BS located at some fixed position (x_0, y_0), *say, the Laplace transform of the interference* V *at the user, i.e. the total received power at the user from all BSs in the deployment other than the serving BS, is given by*

$$\mathbb{E}\left[e^{-sV} \mid \text{serving BS fixed at } (x_0, y_0)\right]$$
$$= \exp\left\{-\pi\lambda P^{2/\alpha}\,\mathbb{E}[H^{2/\alpha}]\,\Gamma\left(1-\frac{2}{\alpha}\right)s^{2/\alpha}\right\}, \quad s > 0,\ \alpha > 2, \tag{4.17}$$

where H *is a random variable with PDF* $f_H(\cdot)$.

Now let us return to the single-tier model (2.4), where the serving BS is located at (x_0, y_0) and the fading attenuation on the link from the serving BS to the user at the origin is denoted H_*. As before, we assume $\{H_b\}_{b\in\Phi'}$ i.i.d. $f_H(\cdot)$. However, the PDF of H_* may in general differ from $f_H(\cdot)$, e.g. if the serving BS performs beamforming. We therefore need to analyze the case where the distribution of H_* is arbitrary, and not necessarily the same as that of H. We do so in the following, starting as usual with the simplest case, where $H_* \sim \text{Exp}(1)$.

Rayleigh fading from serving BS, i.i.d. arbitrary fading on all other links

Suppose that $H_* \sim \text{Exp}(1)$. Then the CCDF of the SINR at the user conditioned on the serving BS being located at (x_0, y_0) is given by (2.5), where $V = \sum_{y\in\tilde{\Phi}'} y$. From the argument in the preceding paragraph, it follows that the Laplace transform of V is given by (4.17). Substituting into (2.5) then gives the following result.

THEOREM 4.5 *Suppose the fade attenuation on the link from the serving BS to the user is exponentially distributed with unit mean and is independent of the fade attenuations on the links to the user from all other BSs. Suppose the latter are i.i.d. and* H *is a random variable with this common distribution. Conditioned on the serving BS being at the fixed location* (x_0, y_0), *the CCDF of the instantaneous SINR* $\varGamma$ *at the user (assumed located at the origin) is given by*

$$\mathbb{P}\{\varGamma > \gamma \mid \text{serving BS at } (x_0, y_0)\}$$
$$= \exp\left\{-sN_0 - \pi\lambda P^{2/\alpha}\,\mathbb{E}[H^{2/\alpha}]\,\Gamma\left(1-\frac{2}{\alpha}\right)s^{2/\alpha}\right\}\Bigg|_{s=\gamma(x_0^2+y_0^2)^{\alpha/2}/P}. \tag{4.18}$$

Note the distinction between the *function* $\Gamma(\cdot)$ and the *random variable* $\varGamma$ in (4.18).

Independent arbitrary fading, i.i.d. on all links except possibly from serving BS

Note that (4.18) was calculated under the assumption of Rayleigh fading on the link between the serving BS and the user. Thus, if H_* in (4.18) is not Exp(1) but has PDF $f_{H_*}(\cdot)$, say, then (4.18) is only the first step in the calculation of $\mathbb{P}\{\varGamma > \gamma \mid \text{serving BS at } (x_0, y_0)\}$, which proceeds as follows.

(1) Write the right side of (4.18) as an analytic function of s:

$$F(s) = \exp\left\{-sN_0 - \pi\Gamma(1-2/\alpha)\,\mathbb{E}[H^{2/\alpha}]\,\lambda(sP)^{2/\alpha}\right\}, \quad s > 0. \tag{4.19}$$

(2) Express the PDF $f_{H_*}(\cdot)$ of H_* as a mixture of Erlang PDFs as in (2.17):

$$f_{H_*}(x) = \sum_{l=1}^{\infty} \beta_l \frac{c_l^{m_l+1}}{m_l!} x^{m_l} \exp(-c_l x), \ x \geq 0, \quad m_l \in \{0, 1, \dots\}, \ \sum_{l=1}^{\infty} \beta_l = 1. \quad (4.20)$$

(3) Use (2.19), with $n = 1$, $\mathbf{A} = 1$, $\boldsymbol{X} = H_*$, $\boldsymbol{b} = \gamma(x_0^2 + y_0^2)^{\alpha/2}/P$, W in (2.19) replaced by $V + N_0$, and $\mathcal{L}_W(s)$ in (2.19) replaced by $F(s)$, to obtain $\mathbb{P}\{\Gamma > \gamma \mid \text{serving BS at } (x_0, y_0)\}$.

THEOREM 4.6 *Conditioned on the serving BS being at the fixed location* (x_0, y_0), *the CCDF of the instantaneous SINR* Γ *at the user (assumed located at the origin) is given by*

$$\mathbb{P}\{\Gamma > \gamma \mid \textit{serving BS at } (x_0, y_0)\}$$
$$= \sum_{l=1}^{\infty} \frac{(-1)^{m_l} \beta_l c_l^{m_l+1}}{m_l!} \frac{\partial^{m_l}}{\partial c_l^{m_l}} c_l^{-1} F\left(\frac{c_l \gamma (x_0^2 + y_0^2)^{\alpha/2}}{P}\right), \quad \gamma > 0, \quad (4.21)$$

where $F(s)$ *is given by* (4.19) *and the PDF of the fade attenuation* H_* *on the link from the serving BS to the user is written as a mixture of Erlang PDFs in the form* (4.20).

In practice, we would replace the infinite sum in (4.20) by some finite number of terms, chosen large enough that the truncated series on the right of (4.20) is an accurate approximation of $f_{H_*}(\cdot)$, perhaps using one of the methods mentioned in Remark 2.4. Note that this is possible because $f_{H_*}(\cdot)$ can be uniformly approximated to any desired accuracy by a mixture of Erlang PDFs. This also causes the infinite sum in (4.21) to be replaced by a finite sum.

In Hunter *et al.* (2008, eqn. (7)), the CCDF $\mathbb{P}\{\Gamma > \gamma \mid \text{serving BS at} (x_0, y_0)\}$ was calculated for i.i.d. Rayleigh fading on all links between BSs and the user, then variously extended to i.i.d. mixture–Nakagami fading, sectored-beam antennas at the BSs, and MIMO transmissions with eigen-beamforming. The same CCDF was also calculated for MIMO transmissions with a minimum mean squared error (MMSE) receiver and i.i.d. Rayleigh fading in Ali, Cardinal & Gagnon (2010, eqn. (8)). An expression for $\mathbb{P}\{\Gamma > \gamma \mid \text{serving BS at } (x_0, y_0)\}$ for i.i.d. Rayleigh fading on all links between BSs and the user and an arbitrary path-loss model instead of the slope-intercept model (1.1) is derived in Zuyev (2010, eqn. (16.2), p. 536).

We now have a full solution of the CCDF of the SINR at a user location for a BS deployment whose locations are modeled by the points of a homogeneous PPP, for arbitrary i.i.d. fading on all links between BSs and the user, *conditioned* on a fixed point of the PPP being the location of the serving BS for this user. To obtain the *unconditional* CCDF of the SINR, we need to average over the distribution of the location of the serving BS relative to the user, which depends upon the criterion for selection of the serving BS.

4.3.2 A note on serving BS selection criteria

In cellular networks, a user is usually served by the BS whose transmissions it receives "most strongly" in some sense. From (2.1), the received power at the user depends not only on distance, but also on the fading attenuation on the link from the BS to the user. Generally, the choice of serving BS is not the one with maximum instantaneous received power, but the one with maximum weighted average value obtained from several successive measurements of instantaneous received power. Assuming the user has not moved, or has moved very little, over the period of these measurements, this averaging has the effect of smoothing out the variations due to fast fading, with the extent of smoothing depending upon the number of measurements being averaged and the method of averaging.

In this chapter, we analyze two serving BS selection criteria, which differ in the extent of smoothing of the instantaneous measurements.

(1) The *nearest-BS association* rule is when the user is served by (equivalently, *associates with*) the BS that is geographically nearest to it. While this does not hold for cellular networks deployed today, this serving BS selection criterion corresponds to the case where the user is static relative to the BSs, and a large number of successive instantaneous received power measurements are averaged so that the effect of fast fading is completely eliminated. Then the average of these measurements is simply the received power assuming no fading, so the serving BS selection is based only on the path loss (1.1).

(2) The *strongest-BS association* rule is when the user is served by, or associates with, the BS that it receives most strongly, i.e. with maximum instantaneous received power among all BSs in the tier. This serving BS selection criterion corresponds to the other extreme of smoothing, namely no smoothing at all. Here the choice of serving BS is made based on the *instantaneous* received signal power (2.1).

Unfortunately for our approach so far, conditioning upon the serving BS selection criterion introduces a dependence between the non-serving BSs and the serving BS. For example, if the serving BS is chosen to be the BS that is nearest to the user location, then, conditioned on this distance being r_0, say, we know that all non-serving BSs must be at a distance $>r_0$ from the user location. In general, Theorem 4.3 does not hold if we condition upon the serving BS selection criterion, so the Laplace transform of V in (2.2) cannot be substituted by the Laplace transform of W in (4.7). Nonetheless, the previous theoretical results on PPPs do allow us to calculate the Laplace transform of W in (2.19) for a scenario where the serving BS for the user is the BS that is geographically nearest to the user location (the origin). We show how to do so in Section 4.3.3.

4.3.3 Serving BS is the one nearest to the user

Consider again the PPP Φ from Section 4.2.1 modeling the locations of BSs in the single-tier deployment. Suppose we have a user at the origin. Our association rule means that this user is served by the point of Φ nearest to the origin. Let the distance of the serving BS from the user (i.e. the origin) be R_1, and suppose we are given $R_1 = r_1$, say. Note that R_1 is a random variable defined by the distance of the nearest point of the PPP Φ from the origin. We show later how to derive the distribution of R_1.

Now, observe that all BSs other than the serving BS must be at a distance greater than r_1 from the user location. Further, all these BSs cause interference at the user. The locations of these BSs are given by the points of a point process Φ_{r_1} obtained by *thinning* the PPP Φ such that a point of Φ at location $(x, y) \in \mathbb{R}^2$ is retained in Φ_{r_1} with retention probability given by

$$p_{r_1}(x,y) = \begin{cases} 1, & \sqrt{x^2+y^2} > r_1, \\ 0, & \text{otherwise}, \end{cases}$$
$$= 1_{(r_1,\infty)}\left(\sqrt{x^2+y^2}\right), \quad (x,y) \in \mathbb{R}^2,$$

where, for any d and any set $\mathcal{A} \subseteq \mathbb{R}^d$, the function $1_{\mathcal{A}} : \mathbb{R}^d \to \{0,1\}$ defined by

$$1_{\mathcal{A}}(\boldsymbol{x}) = \begin{cases} 1, & \boldsymbol{x} \in \mathcal{A}, \\ 0, & \text{otherwise}, \end{cases}$$

is called the *indicator function* of the set $\mathcal{A}$.

We know that the point process Φ_{r_1} is a PPP with intensity function $\lambda_{r_1}(x,y) = p_{r_1}(x,y)\lambda(x,y)$, $(x,y) \in \mathbb{R}^2$, where $\lambda(\cdot,\cdot)$ is the intensity function of Φ. With the function f as defined in (4.1) and following the derivation in Section 4.2.1, it is easy to see that the one-dimensional point process of received powers at the user from the non-serving BSs in the deployment,

$$\tilde{\Phi}_{r_1} = \{y \in \mathbb{R}_+ : y = f(b, H_b), b \in \Phi_{r_1}, H_b \text{ i.i.d. } f_H(\cdot)\},$$

is a one-dimensional PPP with mean measure

$$\tilde{\mu}_{r_1}((0,y]) = \int_0^\infty r \int_0^{2\pi} \lambda_{r_1}(r\cos\theta, r\sin\theta)\mathrm{d}\theta\, \mathrm{d}r \int_0^{yr^\alpha/P} f_H(h)\mathrm{d}h$$
$$= \int_{r_1}^\infty r \int_0^{2\pi} \lambda(r\cos\theta, r\sin\theta)\mathrm{d}\theta\, \mathrm{d}r \int_0^{yr^\alpha/P} f_H(h)\mathrm{d}h, \quad y \geq 0,$$

or, equivalently, intensity function

$$\tilde{\lambda}_{r_1}(y) = \int_{r_1}^\infty \frac{r^{1+\alpha}}{P} f_H\left(\frac{yr^\alpha}{P}\right) \int_0^{2\pi} \lambda(r\cos\theta, r\sin\theta)\mathrm{d}\theta\, \mathrm{d}r, \quad y \geq 0.$$

As before, we assume Φ is homogeneous with constant intensity (density) λ. Then we obtain

$$\tilde{\lambda}_{r_1}(y) = 2\pi\lambda \int_{r_1}^\infty \frac{r^{1+\alpha}}{P} f_H\left(\frac{yr^\alpha}{P}\right) \mathrm{d}r = \frac{2\pi\lambda P^{2/\alpha}}{\alpha y^{1+2/\alpha}} \int_{yr_1^\alpha/P}^\infty t^{2/\alpha} f_H(t)\mathrm{d}t, \quad y \geq 0. \quad (4.22)$$

Conditional interference distribution

Consider again the single-tier model (2.4), where now the serving BS is the one that is nearest to the user location. Suppose also that the fade attenuation on the link from the serving BS is $H_* \sim \text{Exp}(1)$, while the fade attenuations on the links from all other BSs are i.i.d. with arbitrary PDF $f_H(\cdot)$. From (2.4) and (2.5), with R_* replaced by R_1, we see that the CCDF of the SINR Γ at the user location conditioned on the given distance to the serving (i.e. nearest) BS, $R_1 = r_1$, is given by

$$\mathbb{P}\{\Gamma > \gamma \mid R_1 = r_1\} = \exp(-sN_0)\mathcal{L}_{V \mid R_1=r_1}(s)\big|_{s=\gamma r_1^\alpha/P}, \tag{4.23}$$

where, from (2.2), conditioned on $R_1 = r_1$, V is the total interference power at the user, i.e. the total received power at the user from all BSs in the tier other than the serving BS:

$$V = \sum_{y \in \tilde{\Phi}_{r_1}} y.$$

We can prove the following counterpart to Theorem 4.4.

THEOREM 4.7 *Suppose that the fade attenuations on all links to the user from BSs in the deployment are i.i.d. with common PDF $f_H(\cdot)$, and suppose that the user is served by the nearest BS. Conditioned on the distance to this BS being r_1, say, the Laplace transform of the interference V at the user, i.e. the total received power at the user from all BSs in the deployment other than the serving BS, is given by*

$$\mathcal{L}_{V \mid R_1=r_1}(s) = \exp\left[-\pi\lambda(Ps)^{2/\alpha} G_\alpha\left(\frac{r_1^2}{P^{2/\alpha}s^{2/\alpha}}\right)\right], \quad s > 0, \tag{4.24}$$

where

$$G_\alpha(z) = \mathbb{E}\left[\gamma\left(1 - \frac{2}{\alpha}, \frac{H}{z^{\alpha/2}}\right) H^{2/\alpha}\right] - z\left[1 - \mathcal{L}_H\left(\frac{1}{z^{\alpha/2}}\right)\right], \quad z \geq 0, \tag{4.25}$$

H is a random variable with PDF $f_H(\cdot)$, and

$$\gamma(x, y) = \int_0^y t^{x-1}\, \mathrm{e}^{-t}\, \mathrm{d}t, \quad x \geq 0,\, y \geq 0,$$

is the lower incomplete gamma *function.*

Proof The derivation of (4.24) follows similar steps to the derivation of (4.11). Problem 4.5 shows the sequence of algebraic manipulations that yield (4.24) and (4.25). □

Note the distinction between the *function* $\gamma(\cdot, \cdot)$ and the *argument* γ of the CCDF $\mathbb{P}\{\Gamma > \gamma\}$, just as we distinguish between the *function* $\Gamma(\cdot)$ and the *random variable* Γ.

Remark 4.7 An alternative expression to (4.24) for the Laplace transform of V conditioned on $R_1 = r_1$ is derived in Problem 4.6. If the fade attenuations on links to the user from all BSs in Φ are i.i.d. and H is a random variable with this common distribution, an interesting result linking the Laplace transform of V conditioned on $R_1 = r_1$ and the

Laplace transform of W, the total received power at the user from all BSs in Φ (including the nearest one, at distance R_1), may be obtained from the following argument: as $W = PH_*/R_1^\alpha + V$, it follows by conditioning on R_1 that

$$
\begin{aligned}
\mathcal{L}_W(s) &= \mathbb{E}\left[\exp\left(-\frac{PsH_*}{R_1^\alpha} - sV\right)\right] \\
&= \mathbb{E}_{R_1}\left\{\mathbb{E}\left[\exp\left(-\frac{PsH_*}{R_1^\alpha} - sV\right) \,\middle|\, R_1\right]\right\} \\
&= \mathbb{E}_{R_1}\left\{\mathbb{E}_{H_*}\left[\exp\left(-\frac{PsH_*}{R_1^\alpha}\right)\right] \mathbb{E}[\exp(-sV) \mid R_1]\right\} \\
&= \mathbb{E}_{R_1}\left[\mathcal{L}_H\left(\frac{Ps}{R_1^\alpha}\right)\mathcal{L}_{V\mid R_1}(s)\right] = \int_0^\infty f_{R_1}(r_1)\mathcal{L}_H\left(\frac{Ps}{r_1^\alpha}\right)\mathcal{L}_{V\mid R_1=r_1}(s)\mathrm{d}r_1 .
\end{aligned}
\tag{4.26}
$$

A direct proof of this result is outlined in Problem 4.7.

Remark 4.8 Another interesting problem is the Laplace transform of total received power W from all BSs in the tier when the user is known to be at least at a distance of $d_{\min}$ from all BSs in the tier, i.e. when $R_1 > d_{\min}$. Let us denote this Laplace transform by $\mathcal{L}_{W\mid R_1>d_{\min}}(\cdot)$. In Problem 4.8, we show that

$$\mathcal{L}_{W\mid R_1>d_{\min}}(s) = \mathcal{L}_{V\mid R_1=d_{\min}}(s), \quad s > 0, \tag{4.27}$$

where $\mathcal{L}_{V\mid R_1=d_{\min}}(s)$ is given by (4.24).

Conditional SINR distribution

Substituting (4.24) into (4.23), we obtain our first result.

THEOREM 4.8 *Under the nearest-BS association rule, i.i.d. arbitrary fading with PDF $f_H(\cdot)$ on the links from all non-serving BSs to the user location, and when the fade attenuation on the link from the serving BS to the user is exponentially distributed with unit mean, the CCDF of the SINR at the user conditioned on the distance from the serving (i.e. nearest) BS is given by*

$$\mathbb{P}\{\Gamma > \gamma \mid R_1 = r_1\} = \exp\left[-\frac{\gamma r_1^\alpha N_0}{P} - \pi\lambda r_1^2 \gamma^{2/\alpha} G_\alpha\left(\frac{1}{\gamma^{2/\alpha}}\right)\right], \quad \gamma > 0, \tag{4.28}$$

where the function $G_\alpha(\cdot)$ is given by (4.25) *and H is a random variable with PDF $f_H(\cdot)$.*

It is important to observe that (4.28), together with (4.25), is the first SINR distribution result in our study so far to correspond to a "realistic" serving BS selection criterion. Historically, it was also the first such distribution to be derived, and its derivation by Andrews *et al.* (2011) was a significant breakthrough in our understanding of the distribution of the SINR in cellular networks. We see from (4.28) by setting $N_0 = 0$ that *under the nearest-BS association rule, the distribution of the SIR (but not the SINR) conditioned on the distance to the serving BS is not a function of the transmit power of the BSs.*

From (4.24) and (4.25), we obtain the conditional CCDF of the SINR given by (4.28). When $H \sim \text{Exp}(1)$, we can show that (see Problem 4.9)

$$H \sim \text{Exp}(1) \Rightarrow G_\alpha(z) = \begin{cases} \dfrac{2}{\alpha-2}\dfrac{z}{(1+z^{\alpha/2})}\,{}_2F_1\left(1,1;2-\dfrac{2}{\alpha};\dfrac{1}{1+z^{\alpha/2}}\right), & z>0, \\ \dfrac{2\pi/\alpha}{\sin(2\pi/\alpha)} \equiv \dfrac{1}{\text{sinc}(2/\alpha)}, & z=0, \end{cases} \tag{4.29}$$

where ${}_2F_1(a,b;c;z)$ is the *hypergeometric* function,

$${}_2F_1(a,b;c;z) = 1 + \sum_{k=1}^{\infty}\left[\prod_{l=0}^{k-1}\frac{(a+l)(b+l)}{(c+l)}\right]\frac{z^k}{k!}, \quad z \in \mathbb{R}. \tag{4.30}$$

Further, if $\alpha = 4$, then (see Problem 4.10)

$$H \sim \text{Exp}(1) \Rightarrow G_4(z) = \cot^{-1}(z), \quad z \geq 0. \tag{4.31}$$

Again, if H_* is not Exp(1), the full derivation of $\mathbb{P}\{\Gamma > \gamma \mid R_1 = r_1\}$ requires going through steps (i)–(iii) as discussed in Section 4.3.1, with the function $F(s)$ in step (i) replaced by $F_{r_1}(s)$, say, where

$$F_{r_1}(s) = \exp\left\{-sN_0 - \pi\lambda(Ps)^{2/\alpha}\left(\mathbb{E}\left[\gamma\left(1-\frac{2}{\alpha},\frac{sPH}{r_1^\alpha}\right)H^{2/\alpha}\right]\right.\right.$$
$$\left.\left. - \frac{r_1^2}{(Ps)^{2/\alpha}}\left[1-\mathcal{L}_H\left(\frac{sP}{r_1^\alpha}\right)\right]\right)\right\}. \tag{4.32}$$

THEOREM 4.9 *Under the nearest-BS association rule, i.i.d. arbitrary fading with PDF $f_H(\cdot)$ on the links from all non-serving BSs to the user location, and when the fade attenuation H_* on the link from the serving BS to the user is distributed with PDF $f_{H_*}(\cdot)$, the CCDF of the SINR at the user conditioned on the distance from the serving (i.e. nearest) BS is given by*

$$\mathbb{P}\{\Gamma > \gamma \mid R_1 = r_1\} = \sum_{l=1}^{\infty}\frac{(-1)^{m_l}\beta_l c_l^{m_l+1}}{m_l!}\frac{\partial^{m_l}}{\partial c_l^{m_l}}c_l^{-1}F_{r_1}\left(\frac{c_l\gamma r_1^\alpha}{P}\right), \quad \gamma > 0, \tag{4.33}$$

where $F_{r_1}(\cdot)$ is defined by (4.32) *and $f_{H_*}(\cdot)$ is written as a mixture of Erlang PDFs in the form* (4.20).

In practice, the infinite mixture in (4.20) would be approximated by a finite mixture by truncating the infinite series after a finite number of terms, perhaps using one of the methods mentioned in Remark 2.4. This also turns the infinite series in (4.33) into a finite sum.

Unconditional SINR distribution

The unconditional CCDF $\mathbb{P}\{\Gamma > \gamma\}$ can be obtained by integrating over the PDF of R_1, which is given by (Kingman, 1993, eqn. (2.35), p. 21)

$$f_{R_1}(r_1) = 2\pi\lambda r_1 \exp\left(-\pi\lambda r_1^2\right), \quad r_1 \geq 0. \tag{4.34}$$

The distribution of R_1 is an immediate consequence of the definition of a PPP: see Problem 4.11.

We begin with the case where the fade attenuation H_* on the link from the serving BS to the user is exponentially distributed with unit mean. The links from all other BSs to the user are assumed to have i.i.d. arbitrary fading with PDF $f_H(\cdot)$. With the change of variables $u_1 = \pi\lambda r_1^2$, the unconditional CCDF of the SINR is obtained from (4.28) and (4.34).

THEOREM 4.10 *Under the nearest-BS association rule, i.i.d. arbitrary fading with PDF $f_H(\cdot)$ for the fade attenuations on the links from all non-serving BSs to the user location, and when the fade attenuation on the link from the serving BS to the user is exponentially distributed with unit mean, the CCDF of the SINR at the user conditioned on the distance from the serving (i.e. nearest) BS is given by*

$$\mathbb{P}\{\Gamma > \gamma\} = \int_0^\infty \exp\left\{-\frac{\gamma}{\pi^{\alpha/2}} \frac{u_1^\alpha}{\lambda^{\alpha/2} P/N_0} - \left[1 + \gamma^{2/\alpha} G_\alpha\left(\frac{1}{\gamma^{2/\alpha}}\right)\right] u_1\right\} du_1, \quad \gamma > 0, \tag{4.35}$$

where $G_\alpha(\cdot)$ is defined in (4.25) *and H is a random variable with PDF $f_H(\cdot)$.*

Note from (4.35) that *the distribution of the SINR under the nearest-BS association rule depends on the parameters of the deployment only in the form $\lambda^{\alpha/2} K P^{\mathrm{tx}}/N_0$*. Thus the right side of (4.35) could be numerically computed off-line for various choices of the parameter $\lambda^{\alpha/2} K P^{\mathrm{tx}}/N_0$, and a look-up table created that is then applicable to any set of deployment parameters. When thermal noise is negligible ($N_0 = 0$), we can integrate the conditional CCDF (4.28) with respect to the PDF (4.34) to obtain the unconditional CCDF of the SIR in closed form under the nearest-BS association rule:

$$N_0 = 0 \Rightarrow \mathbb{P}\{\Gamma > \gamma\} = \frac{1}{1 + \gamma^{2/\alpha} G_\alpha(\gamma^{-2/\alpha})}, \quad \gamma > 0. \tag{4.36}$$

Thus *the distribution of the SIR (but not the SINR) under the nearest-BS association rule does not depend on the transmit power or the density of the BS deployment.*

When $H \sim \mathrm{Exp}(1)$, i.e. there is i.i.d. Rayleigh fading on all links from all BSs to the user location, and thermal noise power is negligible ($N_0 = 0$), from (4.36) we have

$$N_0 = 0,\ H \sim \mathrm{Exp}(1) \Rightarrow \mathbb{P}\{\Gamma > \gamma\} = \frac{1}{1 + \dfrac{2\rho\ {}_2F_1(1,1;2-2/\alpha;\rho)}{\alpha - 2}}, \quad \rho = \frac{\gamma}{1+\gamma}.$$

Since $0 < \rho < 1$, it follows from (4.30) that the hypergeometric function ${}_2F_1(1,1;2-2/\alpha;\cdot)$ evaluated at ρ is a well-behaved convergent series of terms, all of the same sign. Thus it can be evaluated to any desired accuracy simply by truncating the series after a finite number of terms. Algorithms for efficient numerical computation of ${}_2F_1(a,b;c;\cdot)$ functions are proposed and studied in detail in Pearson (2009), and Matlab® code for the algorithms is also provided.

For the special case of i.i.d. Rayleigh fading, no thermal noise, and path-loss exponent $\alpha = 4$, (4.31) yields the following simple expression for the CCDF of the SINR:

$$\alpha = 4,\ N_0 = 0,\ H \sim \mathrm{Exp}(1) \Rightarrow \mathbb{P}\{\Gamma > \gamma\} = \frac{1}{1 + \sqrt{\gamma}\cot^{-1}\left(\frac{1}{\sqrt{\gamma}}\right)},\ \gamma > 0. \tag{4.37}$$

Similarly, when the fade attenuation H_* on the link from the nearest BS to the user is not exponential and its PDF $f_{H_*}(\cdot)$ is expressed in terms of an infinite mixture of Erlang PDFs of the form (4.20), the change of variables $u_1 = \pi\lambda r_1^2$ and interchanging the order of differentiation and integration in (4.33) yields the following result.

THEOREM 4.11 *Under the nearest-BS association rule, i.i.d. arbitrary fading with PDF $f_H(\cdot)$ on the links from all non-serving BSs to the user location, and when the fade attenuation H_* on the link from the serving BS to the user is distributed with PDF $f_{H_*}(\cdot)$, expressed in terms of an infinite mixture of Erlang PDFs of the form* (4.20), *the CCDF of the SINR at the user is given by*

$$\mathbb{P}\{\Gamma > \gamma\} = \sum_{l=1}^{\infty} \frac{(-1)^{m_l}\beta_l c_l^{m_l+1}}{m_l!} \frac{\partial^{m_l}}{\partial c_l^{m_l}} \frac{1}{c_l} \int_0^{\infty} \exp\Bigg\{ -\frac{c_l\gamma}{\pi^{\alpha/2}} \frac{N_0}{\lambda^{\alpha/2} K P^{\mathrm{tx}}} u_1^{\alpha/2} - u_1\, \mathbb{E}\left[\gamma\left(1 - \frac{2}{\alpha}, c_l\gamma H\right)(c_l\gamma H)^{2/\alpha} + \exp(-c_l\gamma H)\right]\Bigg\}\, du_1,\quad \gamma > 0. \tag{4.38}$$

Proof The proof is by straightforward algebra, the steps of which are outlined in Problem 4.12. □

Remark 4.9 Another scenario of interest is where we want the SINR distribution for a user that is at least a distance of, say, $r_{\min}$ from the nearest BS in Φ. In other words, the lower limit on R_1 is not zero but $r_{\min}$. Equivalently, (4.38) now becomes

$$\mathbb{P}\{\Gamma > \gamma\} = \sum_{l=1}^{\infty} \frac{(-1)^{m_l}\beta_l c_l^{m_l+1}}{m_l!} \frac{\partial^{m_l}}{\partial c_l^{m_l}} \frac{1}{c_l} \int_{u_{\min}}^{\infty} \exp\Bigg\{ -\frac{c_l\gamma}{\pi^{\alpha/2}} \frac{N_0}{\lambda^{\alpha/2} K P^{\mathrm{tx}}} u_1^{\alpha/2} - u_1\, \mathbb{E}\left[\gamma\left(1 - \frac{2}{\alpha}, c_l\gamma H\right)(c_l\gamma H)^{2/\alpha} + \exp(-c_l\gamma H)\right]\Bigg\}\, du_1,\quad \gamma > 0, \tag{4.39}$$

where $u_{\min} = \pi\lambda r_{\min}^2$.

Remark 4.10 From (4.38), it follows that, for arbitrary H_*, the distribution of the SINR under the nearest-BS association rule depends on the parameters of the deployment only in the form $\lambda^{\alpha/2}KP^{\text{tx}}/N_0$. Further, the unconditional CCDF of the SIR can be obtained by setting $N_0 = 0$ to be

$$N_0 = 0 \Rightarrow \mathbb{P}\{\Gamma > \gamma\} = \sum_{l=1}^{\infty} \frac{(-1)^{m_l}\beta_l c_l^{m_l+1}}{m_l!} \times \frac{\partial^{m_l}}{\partial c_l^{m_l}} \left\{ c_l \mathbb{E}\left[\gamma\left(1-\frac{2}{\alpha}, c_l\gamma H\right)(c_l\gamma H)^{2/\alpha} + \exp(-c_l\gamma H)\right]\right\}^{-1}, \quad \gamma > 0.$$

Problems

4.5 Prove (4.24) as outlined in the following.

(1) Use (4.22) and follow similar steps as in the derivation of (4.11) to show that

$$\begin{aligned}
&\mathcal{L}_{V|R_1=r_1}(s) \\
&= \exp\left\{-\int_0^{\infty}\left(1-\mathrm{e}^{-sx}\right)\tilde{\lambda}_{r_1}(x)\mathrm{d}x\right\} \\
&= \exp\left\{-\frac{2\pi\lambda P^{2/\alpha}}{\alpha}\int_0^{\infty}\frac{1-\mathrm{e}^{-sx}}{x^{1+2/\alpha}}\int_{xr_1^{\alpha}/P}^{\infty} t^{2/\alpha} f_H(t)\mathrm{d}t\,\mathrm{d}x\right\} \\
&= \exp\left\{-\frac{2\pi\lambda P^{2/\alpha}}{\alpha}\int_0^{\infty} t^{2/\alpha} f_H(t)\int_0^{Pt/r_1^{\alpha}}\frac{1-\mathrm{e}^{-sx}}{x^{1+2/\alpha}}\,\mathrm{d}x\,\mathrm{d}t\right\},
\end{aligned} \tag{4.40}$$

where in the final step we have switched the order of integration.

(2) Next, use integration by parts, (4.14), and the definition of the incomplete gamma function $\gamma(\cdot,\cdot)$ to show that

$$\frac{2}{\alpha}\int_0^{Pt/r_1^{\alpha}}\frac{1-\mathrm{e}^{-sx}}{x^{1+2/\alpha}}\,\mathrm{d}x = s^{2/\alpha}\left[\gamma\left(1-\frac{2}{\alpha},\frac{sPt}{r_1^{\alpha}}\right) - \frac{r_1^2}{(Ps)^{2/\alpha}}\frac{1-\mathrm{e}^{-sPt/r_1^{\alpha}}}{t^{2/\alpha}}\right]. \tag{4.41}$$

(3) Substitute (4.41) into (4.40) to derive (4.24).

4.6 Prove the following alternative expression to (4.24):

$$\mathcal{L}_{V|R_1=r_1}(s) = \exp\left\{-2\pi\lambda\int_{r_1}^{\infty} r\left[1-\mathcal{L}_H\left(\frac{Ps}{r^{\alpha}}\right)\right]\mathrm{d}r\right\}, \quad s > 0. \tag{4.42}$$

Hint: Follow the same steps as in Problem 4.4, with the function $g(r)$ in the final step changed to $g(r) = \mathcal{L}_H(sP\,1_{(r_1,\infty)}(r)/r^{\alpha})$, $r \geq 0$, and use the fact that, for any $r \geq 0$, $1 - \mathcal{L}_H(sP\,1_{(r_1,\infty)}(r)/r^{\alpha}) \equiv 1_{(r_1,\infty)}(r)[1-\mathcal{L}_H(sP/r^{\alpha})]$.

4.7 Verify (4.26). *Hint*: Use (4.34) and the alternative forms (4.15) and (4.42), as follows.

(1) With the change of variables $u = \pi\lambda r^2$ and $u_1 = \pi\lambda r_1^2$ in (4.42), write

$$\mathcal{L}_{V|R_1=r_1}(s) = \exp[-F_s(u_1)], \quad F_s(u_1) = \int_{u_1}^{\infty}\left[1-\mathcal{L}_H\left(\frac{\pi^{\alpha/2}\lambda^{\alpha/2}Ps}{u^{\alpha/2}}\right)\right]\mathrm{d}u. \tag{4.43}$$

(2) From the definition of $F_s(u_1)$ in (4.43), prove that

$$\mathcal{L}_H\left(\frac{Ps}{r_1^\alpha}\right) = \mathcal{L}_H\left(\frac{\pi^{\alpha/2}\lambda^{\alpha/2}Ps}{u_1^{\alpha/2}}\right) = 1 + F_s'(u_1).$$

(3) From (4.34), prove that

$$f_{R_1}(r_1)\mathrm{d}r_1 = \exp(-u_1)\mathrm{d}u_1, \quad u_1 = \pi\lambda r_1^2. \tag{4.44}$$

(4) Prove, using integration by parts, that

$$\int_0^\infty f_{R_1}(r_1)\mathcal{L}_H\left(\frac{Ps}{r_1^\alpha}\right)\mathcal{L}_{V\,|\,R_1=r_1}(s)\mathrm{d}r_1 = \int_0^\infty \exp\left[-u_1 - F_s(u_1)\right]\left[1 + F'(u_1)\right]\mathrm{d}u_1$$

$$= \int_0^\infty \exp[-u_1 - F_s(u_1)]\mathrm{d}u_1 - \mathrm{e}^{-u_1 - F_s(u_1)}\,\Big|_0^\infty - \int_0^\infty \exp[-u_1 - F_s(u_1)]\mathrm{d}u_1$$

$$= \exp[-F_s(0)] = \mathcal{L}_W(s),$$

where in the final step we have used (4.15) and the definition of $F_s(\cdot)$ in (4.43).

4.8 Prove (4.27). *Hint*: Given $R_1 > d_{\min}$, the argument in Remark 4.7 leading to (4.26) now yields

$$\mathcal{L}_{W\,|\,R_1>d_{\min}}(s) = \int_{d_{\min}}^\infty f_{R_1}(r_1)\mathcal{L}_H\left(\frac{Ps}{r_1^\alpha}\right)\mathcal{L}_{V\,|\,R_1=r_1}(s)\mathrm{d}r_1\,.$$

Follow the same procedure as in the final part of Problem 4.7 to show that

$$\mathcal{L}_{W\,|\,R_1>d_{\min}}(s) = \exp[-F_s(\pi\lambda d_{\min}^2)] = \mathcal{L}_{V\,|\,R_1=d_{\min}}(s).$$

4.9 Prove (4.29) as follows.

(1) Using Gradshteyn & Ryzhik (2000, 6.455.2, p. 651), show that if $H \sim \mathrm{Exp}(1)$, then, for any $z > 0$,

$$\mathbb{E}\left[\gamma\left(1 - \frac{2}{\alpha}, \frac{H}{z^{\alpha/2}}\right)H^{2/\alpha}\right] = \int_0^\infty t^{2/\alpha}\mathrm{e}^{-t}\gamma\left(1 - \frac{2}{\alpha}, \frac{t}{z^{\alpha/2}}\right)\mathrm{d}t$$

$$= \frac{z^{1-\alpha/2}\Gamma(2)}{\left(1 - \frac{2}{\alpha}\right)\left(1 + z^{-\alpha/2}\right)^2} \times {}_2F_1\left(1, 2; 2 - \frac{2}{\alpha}; \frac{z^{-\alpha/2}}{1 + z^{-\alpha/2}}\right)$$

$$= \frac{z}{\left(1 - \frac{2}{\alpha}\right)}\frac{1}{1 + z^{\alpha/2}}\left(1 - \frac{1}{1 + z^{\alpha/2}}\right) \times {}_2F_1\left(1, 2; 2 - \frac{2}{\alpha}; \frac{1}{1 + z^{\alpha/2}}\right).$$

(2) Substitute in (4.25) and use $\mathcal{L}_H(s) = 1/(1+s)$ for $H \sim \mathrm{Exp}(1)$ to show that

$$G_\alpha(z) = \frac{z}{(1 - 2/\alpha)} \times \left\{\frac{1}{1 + z^{\alpha/2}}\left[\left(1 - \frac{1}{1 + z^{\alpha/2}}\right){}_2F_1\left(1, 2; 2 - \frac{2}{\alpha}; \frac{1}{1 + z^{\alpha/2}}\right) - \left(1 - \frac{2}{\alpha}\right)\right]\right\}.$$

(3) Use the facts that ${}_2F_1(a,b;c;z) \equiv {}_2F_1(b,a;c;z)$ and ${}_2F_1(0,b;c;z) \equiv 1$, and Gradshteyn & Ryzhik (2000, 9.137.2, p. 1000) to show that

$$\left(1 - \frac{1}{1+z^{\alpha/2}}\right) {}_2F_1\left(1,2;2-\frac{2}{\alpha};\frac{1}{1+z^{\alpha/2}}\right) - \left(1-\frac{2}{\alpha}\right)$$
$$= \frac{2}{\alpha}\, {}_2F_1\left(1,1;2-\frac{2}{\alpha};\frac{1}{1+z^{\alpha/2}}\right),$$

and conclude the derivation of (4.29) for $z > 0$.

(4) Verify that

$$H \sim \text{Exp}(1) \Rightarrow \mathbb{E}[H^{2/\alpha}] = \Gamma(1+2/\alpha). \tag{4.45}$$

(5) For $z = 0$, note from (4.25) that

$$G_\alpha(0) = \mathbb{E}\left[\Gamma\left(1-\frac{2}{\alpha}\right)H^{2/\alpha}\right] = \Gamma\left(1-\frac{2}{\alpha}\right)\mathbb{E}\left[H^{2/\alpha}\right]$$
$$= \Gamma\left(1-\frac{2}{\alpha}\right)\Gamma\left(1+\frac{2}{\alpha}\right) = \frac{1}{\text{sinc}(2/\alpha)}, \tag{4.46}$$

where we have used the identities in Gradshteyn & Ryzhik (2000, 8.331, 8.334, p. 887).

4.10 Derive (4.31) as follows.

(1) For $\alpha = 4$ and $z > 0$, use the result (Gradshteyn & Ryzhik, 2000, 9.121.14, p. 996)

$${}_2F_1\left(1,1;\frac{3}{2};\sin^2\theta\right) = \frac{\theta}{\sin\theta\,\cos\theta}$$

in (4.29) with $z = \cot\theta$ to obtain (4.31).

(2) For $\alpha = 4$ and $z = 0$, verify from (4.29) that $G_4(0) = \pi/2 = \cot^{-1}(0)$.

4.11 We prove (4.34) as follows: by definition, R_1 is the distance of the nearest point of the PPP Φ with density λ from the origin. In other words, the probability that $R_1 > r_1$ is the probability that there is *no point* of Φ in the disk with radius r_1 centered at the origin. From the definition of the PPP, the number of points of Φ in this disk is a Poisson random variable with mean $\pi\lambda r_1^2$, so we have

$$\mathbb{P}\{R_1 > r_1\} = \exp(-\pi\lambda r_1^2), \quad r_1 > 0,$$

from which (4.34) follows by differentiation.

4.12 Prove (4.38) as follows.

(1) Start with

$$\mathbb{P}\{\Gamma > \gamma\} = \int_0^\infty f_{R_1}(r_1)\, \mathbb{P}\{\Gamma > \gamma \,|\, R_1 = r_1\} \mathrm{d}r_1, \quad \gamma > 0, \tag{4.47}$$

where $\mathbb{P}\{\Gamma > \gamma \,|\, R_1 = r_1\}$ is given by (4.33).

(2) If $s = c_I \gamma r_1^\alpha / P$, show that

$$Ps/r_1^\alpha = c_I\gamma, \quad (Ps)^{2/\alpha}/r_1^2 = (c_I\gamma)^{2/\alpha},$$

and substitute into (4.32) to get

$$F_{r_1}\left(\frac{c_I\gamma r_1^\alpha}{P}\right) = \exp\left\{-\frac{c_I\gamma N_0 u_1^{\alpha/2}}{\pi^{\alpha/2}\lambda^{\alpha/2} K P^{\mathrm{tx}}} - u_1\,\mathbb{E}\left[\gamma\left(1 - \frac{2}{\alpha}, c_I\gamma H\right)(c_I\gamma H)^{2/\alpha}\right] + u_1 - u_1 \mathcal{L}_H(c_I\gamma)\right\}.$$

(3) Change variables from r_1 to u_1, use (4.44), and interchange integration with respect to u_1 and differentiation with respect to c_I in (4.47) to obtain (4.38).

4.3.4 Serving BS is the one received most strongly at the user

In this section, we consider the other extreme case of "averaging" of instantaneous received power measurements by the user, namely no averaging at all. At any instant, we assume that the user is served by the BS whose instantaneous received power at the user is highest. Clearly, this is not practical, but the results on SINR distribution for this scenario yield a bound on the best performance (in terms of SINR distribution) obtainable by fast BS switching.

As before, we represent the single tier of BS locations by the points of a PPP Ψ, assumed homogeneous with density ν, say. Let the received power at the user from BS $b \in \Psi$ be denoted Y_b. The user is served by the BS that it receives the most strongly, i.e. by the BS $b^* = \arg\max_{b\in\Psi} Y_b$. Note that, for any b and $\{Y_{b'}\}_{b'\in\Psi}$, $Y_{b^*} \geq Y_b$ and

$$\sum_{b'\in\Psi\setminus\{b\}} Y_{b'} = \sum_{b'\in\Psi} Y_{b'} - Y_b \geq \sum_{b'\in\Psi} Y_{b'} - Y_{b^*} = \sum_{b'\in\Psi\setminus\{b^*\}} Y_{b'},$$

so b^* is the BS yielding maximum SINR Γ at the user:

$$\Gamma = \frac{\max_{b\in\Psi} Y_b}{\sum_{b'\in\Psi\setminus\{\arg\max_{b\in\Psi} Y_b\}} Y_{b'} + N_0} = \max_{b\in\Psi} \frac{Y_b}{\sum_{b'\in\Psi\setminus\{b\}} Y_{b'} + N_0}. \tag{4.48}$$

In other words, the *strongest-BS association* rule is equivalent to the *maximum SINR association* (max-SINR) rule. Note also that, for a fixed N_0, the max-SINR association rule is equivalent to the *max-SIR* association rule.

Distribution of distance to serving BS

LEMMA 4.12 *(Madhusudhanan et al., 2012a.) The PDF of the distance R_* from an arbitrary user to the BS that it receives with maximum power, for a network of BSs located at points of a homogeneous PPP Ψ with density ν and arbitrary i.i.d. fading, with fade attenuations having PDF $f_H(\cdot)$, on all links to the user, is given by*

$$f_{R_*}(r) = 2\pi\nu\,\mathbb{E}[H^{2/\alpha}]r\exp\left(-\pi\nu\,\mathbb{E}[H^{2/\alpha}]r^2\right), \quad r \geq 0. \tag{4.49}$$

Proof From Theorem 4.1, the one-dimensional point process of received powers at the user from all the BSs in Ψ is a PPP equivalent to the one-dimensional PPP of received powers at the user from all BSs in a different homogeneous PPP Φ with density $\lambda = \nu\,\mathbb{E}[H^{2/\alpha}]$, transmitting with the same fixed power P, but with no fading on any link to the user. As there is no fading in Φ, the BS in Φ that is received most strongly at the user is simply the nearest BS, and the distance to the nearest BS in Φ has the PDF (4.34) with $\lambda = \nu\,\mathbb{E}[H^{2/\alpha}]$, which yields (4.49). □

Remark 4.11 Note that the PDF (4.49) does not depend on the transmit power P of the BSs. This is to be expected, because all received powers at the user location are scaled by the same transmit power P.

Remark 4.12 If there is fading, the most strongly received BS at the user is not in general the nearest BS to the user. Thus, if there is fading, the distance from the user to a BS that is not the strongest is not in general $> R_*$, the distance to the strongest BS.

Conditional interference distribution

We can also prove a counterpart to Theorem 4.4 and Theorem 4.7.

THEOREM 4.13 *Suppose all links to the user from all BSs in the deployment are i.i.d. with common PDF $f_H(\cdot)$, and suppose the user is served by the BS b^* that has maximum received power at the user. Then, conditioned on this received power from the serving BS being y^*, say, the Laplace transform of the interference V at the user, i.e. the total received power at the user from all BSs in the deployment other than the serving BS, is given by*

$$\mathcal{L}_{V\,|\,Y_{b^*}=y^*}(s)$$

$$= \exp\left\{\int_0^{y^*}\left[1-\exp(-s\,y)\right]\tilde{\lambda}(y)\,\mathrm{d}y\right\} \tag{4.50}$$

$$= \exp\left\{-\frac{\pi\nu P^{2/\alpha}\,\mathbb{E}[H^{2/\alpha}]}{(y^*)^{2/\alpha}}\left[(sy^*)^{2/\alpha}\gamma\left(1-\frac{2}{\alpha}, sy^*\right)+\mathrm{e}^{-sy^*}-1\right]\right\}, \quad s>0, \tag{4.51}$$

where $\tilde{\lambda}(\cdot)$ is given by

$$\tilde{\lambda}(y) = \frac{2\pi\nu P^{2/\alpha}\,\mathbb{E}[H^{2/\alpha}]}{\alpha y^{1+2/\alpha}}, \quad y \geq 0, \tag{4.52}$$

and H is a random variable with PDF $f_H(\cdot)$.

Proof Conditioned on $Y_{b^*} = y^*$, we have

$$V = \sum_{b \in \Psi \setminus \{b^*\}} Y_b = \sum_{b \in \Psi} Y_b 1_{(0,y^*)}(Y_b) = \sum_{y \in \tilde{\Psi}} y 1_{(0,y^*)}(y),$$

where $\tilde{\Psi} = \{Y_b : b \in \Psi\}$ is a one-dimensional PPP with intensity function given (from (4.6)) by (4.52). Then (4.50) follows from (4.8) with $f(y) = y1_{(0,y^*)}(y)$. The derivation of (4.51) from (4.50) follows the same steps as in the derivation of (4.24), and the algebraic manipulations are outlined in Problem 4.13. □

Remark 4.13 Theorem 4.13 is equivalent to Błaszczyszyn *et al.* (2013, Prop. 7). Unfortunately, it is not as useful in explicitly calculating SINR distributions as Theorem 4.4 and Theorem 4.7, because the result in (4.50) is conditioned on the value of the received strongest power $Y_{b^*} = PH_*/R_*^{\alpha}$, and not on either the value of the distance R_* to the serving BS b^* or the value of the fade attenuation H_* on the link to the user from this serving BS. In Dhillon *et al.* (2011) and Błaszczyszyn *et al.* (2013), the joint distribution of (Y_{b^*}, V) is derived (indirectly in Błaszczyszyn *et al.* (2013)) then used to obtain the Laplace transform of the SINR (the Laplace transform of the reciprocal of the SINR, in the case of Błaszczyszyn *et al.* (2013)), and a numerical algorithm to invert this Laplace transform is provided. However, it is possible to derive an explicit expression for the SINR distribution without requiring inversion of a Laplace transform, as we shall show in the remainder of this section.

Why the i.i.d. Rayleigh fading scenario is also the general case

Note that the fade attenuations on the links from the BSs to the user, assumed i.i.d., can have arbitrary distribution. The following corollary to Theorem 4.1 shows that, for calculation of the distribution of the SINR at the user under the max-SINR association rule, it is sufficient to restrict ourselves to the case where the fade attenuations are i.i.d. Exp(1).

COROLLARY 4.14 *The distribution of the maximum SINR at a user from BSs located at points of a homogeneous PPP Ψ with density ν and arbitrary i.i.d. fading, with fade attenuations having PDF $f_H(\cdot)$, on all links to the user, is the same as the distribution of the maximum SINR at the user from BSs located at points of a homogeneous PPP Φ with density*

$$\lambda = \nu \frac{\mathbb{E}[H^{2/\alpha}]}{\Gamma(1 + 2/\alpha)} \tag{4.53}$$

and i.i.d. Rayleigh fading (i.e. the fade attenuation has Exp(1) distribution) on all links to the user, where H is a random variable with PDF $f_H(\cdot)$. The thermal noise power is assumed to be the same at the user receiver in both models, as are the slope-intercept path-loss parameters and the transmit powers of all BSs.

Proof From Theorem 4.1 and (4.45), it follows that the set of received signal powers at the user location from the BSs located at points of the PPPs Φ and Ψ are both one-dimensional PPPs that are equivalent to the one-dimensional PPP of signal powers received from the BSs located at the points of a homogeneous PPP with density

$\lambda\Gamma(1+2/\alpha) = \nu\,\mathbb{E}[H^{2/\alpha}]$ and no fading on the links. Since the path-loss parameters, transmit power, and thermal noise remain the same for the two transmission scenarios Ψ and Φ, it follows that the distribution of the maximum SINR at the user is also the same for both scenarios. □

Remark 4.14 It is clear that, in general, any fading distribution could be used instead of the Rayleigh distribution for the equivalent BS deployment for analysis, so long as the density of the equivalent deployment is scaled accordingly (see (4.53)). This result was also derived in a different way directly for the distribution of the reciprocal of the SINR when receiving from the strongest BS (Błaszczyszyn *et al.*, 2010, Prop. 5.5). We work with the Rayleigh distribution for fading on the links to the user from the BSs in the network because of its analytical advantages, as seen in Theorem 2.4. An alternative is to work with the equivalent BS deployment with no fading on the links to the user. This approach can be used to yield the characteristic function of the maximum SIR, which then needs to be inverted numerically to obtain the CCDF (Madhusudhanan *et al.*, 2011, Thm. 1). As already stated in Remark 4.13, in this section we show how to derive an explicit expression for the SINR distribution without requiring such numerical inversion of a Fourier or Laplace transform.

CCDF of the SINR from the countable inclusion-exclusion formula

We may therefore focus, without loss of generality, on a deployment where the BSs are located at the points of a homogeneous PPP Φ with density λ, say, and i.i.d. Rayleigh fading on all links between BSs and the user, such that the received power at the user from any BS $b \in \Phi$ at distance R_b and transmitting with power P^{tx} is

$$Y_b = \frac{PH_b}{R_b^{\alpha}}, \; H_b \sim \mathrm{Exp}(1), \; P = KP^{\mathrm{tx}}, \quad b \in \Phi,$$

where (α, K) are, respectively, the slope and intercept of the slope-intercept path-loss model (1.1). For ease of notation, suppose we label the points of Φ as $\Phi = \{b_1, b_2, \ldots\}$. From (4.48), it follows that we may write

$$\mathbb{P}\{\Gamma > \gamma\} = \mathbb{P}\left(\bigcup_{n=1}^{\infty}\left\{\frac{Y_{b_n}}{\sum_{b'\in\Phi\setminus\{b_n\}} Y_{b'} + N_0} > \gamma\right\}\right). \tag{4.54}$$

Now we apply the generalization of the inclusion-exclusion principle to the countable union in (4.54). This yields

$$\mathbb{P}\{\Gamma > \gamma\} = \sum_{n=1}^{\infty}(-1)^{n-1}S_n, \tag{4.55}$$

provided (Friedland & Krop, 2006) the sequence $\{S_n\}_{n=1}^{\infty}$ defined by

$$\begin{aligned} S_n &= \sum_{\substack{i_1,\ldots,i_n:\\ 1\le i_1<\cdots<i_n}} \mathbb{P}\left(\bigcap_{m=1}^{n}\left\{\frac{Y_{b_{i_m}}}{\sum_{b'\in\Phi\setminus\{b_{i_m}\}} Y_{b'} + N_0} > \gamma\right\}\right) \\ &= \frac{1}{n!}\sum_{b_{i_1},\ldots,b_{i_n}} \mathbb{P}\left(\bigcap_{m=1}^{n}\left\{\frac{Y_{b_{i_m}}}{\sum_{b'\in\Phi\setminus\{b_{i_m}\}} Y_{b'} + N_0} > \gamma\right\}\right) \end{aligned} \tag{4.56}$$

converges exponentially to zero: $\limsup_{n\to\infty} S_n^{1/n} < 1$. We shall see later that this condition is trivially satisfied.

Next, observe that the set of received powers $\tilde{\Phi} = \{Y_b : b \in \Phi\}$ is a one-dimensional inhomogeneous PPP with intensity function given by (4.4):

$$\tilde{\lambda}(y) = 2\pi\lambda \int_0^\infty \frac{r^{1+\alpha}}{P} \exp\left(-\frac{yr^\alpha}{P}\right) \mathrm{d}r, \quad y \geq 0. \tag{4.57}$$

Note that we used the more general form (4.4) instead of the simpler form (4.6) with $\mathbb{E}[H^{2/\alpha}] = \Gamma(1 + 2/\alpha)$ because we want to retain the exponential functions. As we shall see, this allows us to apply Theorem 2.4. Before we can do so, however, we need to establish some definitions.

For any $n \geq 1$, define the following function h_n of $\boldsymbol{x} = [x_1, \ldots, x_n]^\top \in \mathbb{R}^n$ and a countable subset $\mathcal{A} \subseteq \mathbb{R}_+$:

$$\begin{aligned} h_n(\boldsymbol{x}, \mathcal{A}) &\equiv h_n(x_1, \ldots, x_n, \mathcal{A}) \\ &= \prod_{m=1}^n 1_{(\gamma,\infty)}\left(\frac{x_m}{\sum_{\substack{l=1\\l\neq m}}^n x_l + \sum_{y\in\mathcal{A}} y + N_0}\right) \qquad (4.58) \\ &= \prod_{m=1}^n 1_{\left(\sum_{y\in\mathcal{A}} y + N_0, \infty\right)}\left(\frac{1}{\rho}x_m - \mathbf{1}_n^\top \boldsymbol{x}\right), \quad \rho = \frac{\gamma}{1+\gamma}, \\ &= 1_{\mathcal{U}_n\left(\sum_{y\in\mathcal{A}} y + N_0\right)}(\boldsymbol{x}), \qquad (4.59) \end{aligned}$$

where

$$\mathcal{U}_n(w) = \{\boldsymbol{x} \in \mathbb{R}^n : \mathbf{A}_n \boldsymbol{x} > w\mathbf{1}_n\},\ w > 0, \quad \mathbf{A}_n = \rho^{-1}\mathbf{I}_n - \mathbf{1}_n\mathbf{1}_n^\top. \tag{4.60}$$

Using (4.56) and the preceding definition of h_n, we can write (4.55) in the form

$$\mathbb{P}\{\Gamma > \gamma\} = \sum_{n=1}^\infty \frac{(-1)^{n-1}}{n!} \mathbb{E}\Bigg[\sum_{\substack{y_1,\ldots,y_n\in\tilde{\Phi}:\\ \forall i\neq j, y_i\neq y_j}} h_n\left(y_1, \ldots, y_n, \tilde{\Phi} \setminus \{y_1, \ldots, y_n\}\right)\Bigg]. \tag{4.61}$$

Applying the extended Slivnyak–Mecke theorem

Next, we use a remarkable result called the *extended Slivnyak–Mecke theorem*, which expresses the expected value of the multiple summation above as an integral of a function over the space defined by the arguments of the summation.

THEOREM 4.15 *(Møller & Waagepetersen, 2004, Thm. 3.3, p. 22.) For a one-dimensional PPP Ψ with intensity function ν and a function $g(x_1, \ldots, x_n, \mathcal{A})$ of $(x_1, \ldots, x_n) \in \mathbb{R}^n$ and a countable subset $\mathcal{A} \subseteq \mathbb{R}$,*

$$\begin{aligned} &\mathbb{E}\Bigg[\sum_{\substack{y_1,\ldots,y_n\in\Psi:\\ \forall i\neq j,\ y_i\neq y_j}} g\left(y_1, \ldots, y_n, \Psi \setminus \{y_1, \ldots, y_n\}\right)\Bigg] \\ &= \int_{\mathbb{R}} \cdots \int_{\mathbb{R}} \mathbb{E}[g(x_1, \ldots, x_n, \Psi)]\nu(x_1)\cdots\nu(x_n)\,\mathrm{d}x_1 \cdots \mathrm{d}x_n. \end{aligned} \tag{4.62}$$

Remark 4.15 The key to the proof is the case $n = 1$ (called the *Slivnyak–Mecke theorem*), as the result for arbitrary $n > 1$ follows from mathematical induction. We do not prove Theorem 4.15 for $n = 1$, but will note here that it is a consequence of *Slivnyak's theorem* (Theorem 4.3). We have only given the statement of the extended Slivnyak–Mecke theorem for a one-dimensional PPP Ψ, but the result is true in general if Ψ is a PPP in $\mathbb{R}^d$ for any d. Note that $x_1, \ldots, x_n$ on the right of (4.62) are just dummy variables of integration. The quantity $g(x_1, \ldots, x_n, \Psi)$ is a random variable because Ψ is a random set, and the expectation of $g(x_1, \ldots, x_n, \Psi)$ on the right of (4.62) is with respect to Ψ.

Remark 4.16 Theorem 4.15 is different from a similar result called Campbell's theorem (Kingman, 1993, p. 28), which states that if Ψ is a PPP on $\mathbb{R}^d$ with intensity function $\nu(\cdot)$ and $f\colon \mathbb{R}^d \to \mathbb{R}_+$, then

$$\mathbb{E}\Big[\sum_{x \in \Psi} f(x)\Big] = \int_{\mathbb{R}} \mathbb{E}[f(x')]\nu(x')\,\mathrm{d}x' . \tag{4.63}$$

Note that (4.63) follows directly from the expression (4.8) for the Laplace transform of $\sum_{x\in\Psi} f(x)$ and the fact that, for any random variable X, $\mathbb{E}[X] = -\frac{\mathrm{d}}{\mathrm{d}s}\mathcal{L}_X(s)|_{s=0}$, which is why Lemma 4.2 is sometimes referred to as Campbell's theorem.

Remark 4.17 Even more confusingly, Theorem 4.15 is sometimes called the Campbell–Mecke theorem (Baddeley, 2007, Thm. 3.2).

For our case, $\Psi = \tilde{\Phi}$ and $\nu(\cdot) = \tilde{\lambda}(\cdot)$ in the statement of the theorem. For $g = h_n$ as defined in (4.59), we have

$$h_n(\boldsymbol{x}, \tilde{\Phi}) = 1_{\mathcal{U}_n(W)}(\boldsymbol{x}),$$

where $\mathcal{U}_n(\cdot)$ is defined in (4.60) and W is the total received power at the user from *all* BSs in Φ, plus the thermal noise power:

$$W = N_0 + \sum_{y \in \tilde{\Phi}} y.$$

Thus the expectation over $\tilde{\Phi}$ reduces to an expectation over W:

$$\mathbb{E}[h_n(\boldsymbol{x}, \tilde{\Phi})] = \mathbb{E}_W\left[1_{\mathcal{U}_n(W)}(\boldsymbol{x})\right],$$

where we explicitly write W as a subscript to the expectation operator to emphasize that the expectation is over the distribution of W. Further, $\tilde{\Phi}$ is a one-dimensional PPP with all its points lying in $\mathbb{R}_{++}$ with probability 1, so the range of integration in each of the integrals on the right side of (4.62) is $\mathbb{R}_{++}$. Using (4.57), let us also define the function

$$\begin{aligned} f_n(\boldsymbol{x}) &= \prod_{m=1}^{n} \tilde{\lambda}(x_m) \\ &= \left(\frac{2\pi\lambda}{P}\right)^n \int_0^\infty \cdots \int_0^\infty \left[\prod_{m=1}^{n} r_m^{1+\alpha}\right] \exp\left(-\frac{1}{P}\left[r_1^\alpha, \ldots, r_n^\alpha\right]^\top \boldsymbol{x}\right) \mathrm{d}r_1 \cdots \mathrm{d}r_n . \end{aligned}$$

Then we can write

$$\mathbb{E}\Bigg[\sum_{\substack{y_{i_1},\ldots,y_{i_n}\in\tilde{\Phi}:\\ \forall j\neq k, y_{i_j}\neq y_{i_k}}} h_n\left(y_{i_1},\ldots,y_{i_n},\tilde{\Phi}\setminus\{y_{i_1},\ldots,y_{i_n}\}\right)\Bigg] \tag{4.64}$$

$$= \int_{\mathbb{R}_{++}}\cdots\int_{\mathbb{R}_{++}} \mathbb{E}[h_n(x_1,\ldots,x_n,\tilde{\Phi})]\tilde{\lambda}(x_1)\cdots\tilde{\lambda}(x_n)\,\mathrm{d}x_1\cdots\mathrm{d}x_n$$

$$= \int_{\mathbb{R}^n_{++}} \mathbb{E}_W\left[1_{\mathcal{U}_n(W)}(\boldsymbol{x})\right] f_n(\boldsymbol{x})\,\mathrm{d}\boldsymbol{x}$$

$$= \mathbb{E}_W\left[\int_{\mathbb{R}^n_{++}} 1_{\mathcal{U}_n(W)}(\boldsymbol{x}) f_n(\boldsymbol{x})\,\mathrm{d}\boldsymbol{x}\right] \tag{4.65}$$

$$= \mathbb{E}_W\left[\int_{\mathcal{U}_n(W)\cap\mathbb{R}^n_{++}} f_n(\boldsymbol{x})\,\mathrm{d}\boldsymbol{x}\right]$$

$$= \left(\frac{2\pi\lambda}{P}\right)^n \mathbb{E}_W\Bigg[\int_{\mathcal{U}_n(W)\cap\mathbb{R}^n_{++}} \int\cdots\int_{\mathbb{R}^n_{++}} \left(\prod_{m=1}^n r_m^{1+\alpha}\right)\exp\left(-\frac{1}{P}\left[r_1^\alpha,\ldots,r_n^\alpha\right]^\top \boldsymbol{x}\right) \times \mathrm{d}r_1\cdots\mathrm{d}r_n\,\mathrm{d}\boldsymbol{x}\Bigg]$$

$$= \left(\frac{2\pi\lambda}{P}\right)^n \int\cdots\int_{\mathbb{R}^n_{++}} \left(\prod_{m=1}^n r_m^{1+\alpha}\right)\mathbb{E}_W\left[\int_{\mathcal{U}_n(W)\cap\mathbb{R}^n_{++}} \exp\left(-\frac{1}{P}\left[r_1^\alpha,\ldots,r_n^\alpha\right]^\top \boldsymbol{x}\right)\mathrm{d}\boldsymbol{x}\right] \times \mathrm{d}r_1\cdots\mathrm{d}r_n, \tag{4.66}$$

where in (4.65) we have interchanged the order of integration and expectation, and have done so again in (4.66).

Given $r_1,\ldots,r_n$, let us now *define* random variables $X_1(r_1),\ldots,X_n(r_n)$, independent of each other and of the random variable W, with the following distributions:

$$X_m(r_m) \sim \mathrm{Exp}\left(\frac{P}{r_m^\alpha}\right), \quad f_{X_m(r_m)}(x) = \frac{r_m^\alpha}{P}\exp\left(-\frac{r_m^\alpha x}{P}\right), \ x\geq 0,\ m=1,\ldots,n.$$

Remark 4.18 It is important to realize that the random variables $X_1(r_1),\ldots,X_n(r_n)$ are entirely fictitious and have been defined only to provide a probabilistic interpretation of the integral in square brackets in (4.66) (see (4.67) in the following). Further, while $X_m(r_m)$ has the *same distribution* as the received power from a BS at distance r_m from the user, we do *not* interpret it as such. Note also that the random variables $X_1(r_1),\ldots,X_n(r_n)$ are independent of W by construction. This provides an elegant solution to the theoretical difficulties (mentioned in the final paragraph of Section 2.3.3) caused by the dependence between the "interference" and the "received power" when the received power comes from the strongest BS.

Define $\boldsymbol{X}(r_1,\ldots,r_n) = [X_1(r_1),\ldots,X_n(r_n)]^\top$. Then $\boldsymbol{X}(r_1,\ldots,r_n)$ has the PDF

$$f_{\boldsymbol{X}(r_1,\ldots,r_n)}(\boldsymbol{x}) = \prod_{m=1}^{n} f_{X_m(r_m)}(x_m) = \left(\prod_{m=1}^{n} \frac{r_m^\alpha}{P}\right) \exp\left(-\frac{1}{P}\left[r_1^\alpha,\ldots,r_n^\alpha\right]^\top \boldsymbol{x}\right),$$
$$\boldsymbol{x} = [x_1,\ldots,x_n]^\top.$$

From the independence of $\boldsymbol{X}(r_1,\ldots,r_n)$ and W, and (4.60), we have

$$\begin{aligned}
&\left(\prod_{m=1}^{n} \frac{r_m^\alpha}{P}\right) \mathbb{E}_W\left[\int_{\mathcal{U}_n(W)\cap\mathbb{R}_{++}^n} \exp\left(-\frac{1}{P}\left[r_1^\alpha,\ldots,r_n^\alpha\right]^\top \boldsymbol{x}\right) \mathrm{d}\boldsymbol{x}\right] \\
&= \mathbb{E}_W\left[\int_{\mathcal{U}_n(W)\cap\mathbb{R}_{++}^n} f_{\boldsymbol{X}(r_1,\ldots,r_n)}(\boldsymbol{x})\,\mathrm{d}\boldsymbol{x}\right] \\
&= \mathbb{E}_W[\mathbb{P}\{\boldsymbol{X}(r_1,\ldots,r_n) \in \mathcal{U}_n(W)\cap\mathbb{R}_{++}^n \mid W\}] \qquad (4.67)\\
&= \mathbb{E}_W[\mathbb{P}\{\mathbf{A}_n\boldsymbol{X}(r_1,\ldots,r_n) > W\mathbf{1}_n \mid W\}] \\
&= \mathbb{P}\{\mathbf{A}_n\boldsymbol{X}(r_1,\ldots,r_n) > W\mathbf{1}_n\} \\
&= \begin{cases} 0, & n\rho \geq 1, \\ \det \mathbf{A}_n^{-1}\left\{\prod_{m=1}^{n} \dfrac{r_m^\alpha}{\left[r_1^\alpha,\ldots,r_n^\alpha\right]^\top \mathbf{A}_n^{-1}\boldsymbol{e}_m^{(n)}}\right\} \mathcal{L}_W\left(\frac{1}{P}\left[r_1^\alpha,\ldots,r_n^\alpha\right]^\top \mathbf{A}_n^{-1}\mathbf{1}_n\right), & n\rho < 1, \end{cases}
\end{aligned} \tag{4.68}$$

where in the final step we use Theorem 2.4 and Problems 2.6 and 2.7. Since $\rho = \gamma/(1+\gamma)$, it follows that (see Problem 4.14)

$$n\rho < 1 \Leftrightarrow n \leq \lceil \gamma^{-1} \rceil, \quad \gamma > 0, \tag{4.69}$$

where, for any $x \in \mathbb{R}$, $\lceil x \rceil$ is the smallest integer greater than or equal to x. From the definition of $\mathbf{A}_n$ in (4.60), (2.23) and (2.22) yield

$$\mathbf{A}_n^{-1} = \rho\left(\mathbf{I}_n + \frac{\rho}{1-n\rho}\mathbf{1}_n\mathbf{1}_n^\top\right), \quad \det \mathbf{A}_n^{-1} = \frac{\rho^n}{1-n\rho}.$$

Further, from (4.11), (4.45), and (4.46), we have

$$\mathcal{L}_W(s) = \exp\left[-sN_0 - \pi\lambda\frac{(Ps)^{2/\alpha}}{\operatorname{sinc}(2/\alpha)}\right], \quad s > 0. \tag{4.70}$$

Distribution of SINR

Substituting (4.70) into (4.68), and thence into (4.66) and finally into (4.61), we obtain the CCDF of the SINR under the max-SINR BS association rule and arbitrary i.i.d. fading to be

$$\mathbb{P}\{\Gamma > \gamma\} = \sum_{n=1}^{\lceil \gamma^{-1} \rceil} \frac{(-1)^{n-1}}{(1-n\rho)\,n!} \left(\frac{1-n\rho}{\rho}\right)^n \int_0^\infty \cdots \int_0^\infty \prod_{m=1}^{n} \frac{2\pi\lambda r_m}{\frac{1-n\rho}{\rho} + \frac{\sum_{k=1}^n r_k^\alpha}{r_m^\alpha}}$$
$$\times \exp\left[-\frac{\rho N_0}{P(1-n\rho)}\left(\sum_{k=1}^n r_k^\alpha\right) - \frac{\pi\lambda}{\mathrm{sinc}(2/\alpha)}\left(\frac{\rho}{1-n\rho}\right)^{2/\alpha}\left(\sum_{k=1}^n r_k^\alpha\right)^{2/\alpha}\right]$$
$$\times\, \mathrm{d}r_1 \cdots \mathrm{d}r_n\,. \tag{4.71}$$

The change of variables

$$u_m = \left(\frac{\rho}{1-n\rho}\right)^{2/\alpha} \pi\lambda r_m^2, \qquad m = 1, \ldots, n,$$

then yields the following result.

THEOREM 4.16 *Suppose we are given BSs whose locations are the points of a homogeneous PPP with density ν. Under the max-SINR BS association rule and i.i.d. arbitrary fading on all links, distributed as some random variable H, the CCDF of the SINR at an arbitrarily located user is given by*

$$\mathbb{P}\{\Gamma > \gamma\} = \sum_{n=1}^{\lceil \gamma^{-1} \rceil} \frac{(-1)^{n-1}}{n!} \frac{1+\gamma}{1-(n-1)\gamma}\left[\frac{1}{\gamma} - (n-1)\right]^{n(1+2/\alpha)}$$
$$\times \int_0^\infty \cdots \int_0^\infty \left\{\prod_{m=1}^n \left[\frac{1}{\gamma} - (n-1) + \frac{\sum_{k=1}^n u_k^{\alpha/2}}{u_m^{\alpha/2}}\right]^{-1}\right\}$$
$$\times \exp\left[-\frac{N_0}{(\pi\lambda)^{\alpha/2} P}\left(\sum_{k=1}^n u_k^{\alpha/2}\right) - \frac{\left(\sum_{k=1}^n u_k^{\alpha/2}\right)^{2/\alpha}}{\mathrm{sinc}(2/\alpha)}\right] \mathrm{d}u_1 \cdots \mathrm{d}u_n,$$
$$\gamma > 0, \tag{4.72}$$

where $\lambda = \nu\,\mathbb{E}[H^{2/\alpha}]/\,\Gamma(1+2/\alpha)$.

As there are only $\lceil \gamma^{-1} \rceil$ terms on the right side of (4.72), it is clear that the condition $\limsup_{n\to\infty} S_n^{1/n} < 1$ is satisfied for the terms defined in (4.56), and so the use of the infinite inclusion-exclusion formula is valid to evaluate the CCDF.

Remark 4.19 From (4.72), we see that the CCDF of the SINR *for arbitrary i.i.d. fading*, evaluated at γ, is given by a sum of $\lceil \gamma^{-1} \rceil$ terms, where the nth term is an n-dimensional integral.

Remark 4.20 Again, we see that *the distribution of the SINR under the max-SIR BS association rule depends on the parameters of the BS deployment only in the combination* $\lambda^{\alpha/2} K P^{\mathrm{tx}}/N_0$, exactly as for the nearest-BS association case. Thus the right side of (4.72) could be numerically computed off-line for various choices of the parameter $\lambda^{\alpha/2} K P^{\mathrm{tx}}/N_0$, and a look-up table created that is then applicable to any set of deployment parameters.

Remark 4.21 Further, setting $N_0 = 0$ in the above equation shows that *the distribution of the SIR (but not the SINR) under the max-SIR BS association rule does not depend on the transmit power or the density of the BS deployment,* also exactly as for the nearest-BS association case. An elegant proof of this result for the SIR under max-SIR BS association is derived from general principles in Madhusudhanan *et al.* (2012b, Cor. 2).

Remark 4.22 Another scenario of interest may be one where we want the distribution of the SINR at a user that is specified to be at least at a distance of, say, $d_{\min}$ from the nearest BS (and therefore all BSs) in Φ. In this case, the lower limit of integration in (4.57) is not zero but $d_{\min}$, and this is carried through (4.66) to (4.71). Equivalently, (4.72) now changes to

$$\mathbb{P}\{\Gamma > \gamma\} = \sum_{n=1}^{\lceil \gamma^{-1} \rceil} \frac{(-1)^{n-1}}{n!} \frac{1+\gamma}{1-(n-1)\gamma} \left[\frac{1}{\gamma} - (n-1)\right]^{n(1+2/\alpha)}$$
$$\times \int_{u_{\min}}^{\infty} \cdots \int_{u_{\min}}^{\infty} \left\{ \prod_{m=1}^{n} \left[\frac{1}{\gamma} - (n-1) + \frac{\sum_{k=1}^{n} u_k^{\alpha/2}}{u_m^{\alpha/2}} \right]^{-1} \right\}$$
$$\times \exp\left[-\frac{N_0}{(\pi\lambda)^{\alpha/2} P} \left(\sum_{k=1}^{n} u_k^{\alpha/2} \right) - \frac{\left(\sum_{k=1}^{n} u_k^{\alpha/2}\right)^{2/\alpha}}{\operatorname{sinc}(2/\alpha)} \right] \mathrm{d}u_1 \cdots \mathrm{d}u_n, \quad \gamma > 0, \tag{4.73}$$

where $u_{\min} = \pi\lambda d_{\min}^2$. Note from (4.4) that Theorem 4.1 and Corollary 4.14 no longer hold with this non-zero minimum-distance requirement, hence (4.73) is only applicable to i.i.d. Rayleigh fading on all links.

Distribution of SIR ($N_0 = 0$)

When $N_0 = 0$, it can be shown (see Problem 4.15) that

$$\mathbb{P}\left\{\Gamma > \gamma \equiv \frac{\rho}{1-\rho}\right\}$$
$$= \sum_{n=1}^{\lceil \gamma^{-1} \rceil \equiv \lceil \rho^{-1} \rceil - 1} \frac{(-1)^{n-1} [\operatorname{sinc}(2/\alpha)]^n}{n(1-n\rho)} \left(\frac{1}{\rho} - n\right)^{2n/\alpha} T_n(\rho, \alpha), \quad 0 < \rho < 1, \tag{4.74}$$

where

$$T_n(\rho,\alpha) = \left(\frac{2}{\alpha}\right)^{n-1} \int_0^1 \mathrm{d}v_1 \int_0^{1-v_1} \mathrm{d}v_2 \cdots \int_0^{1-\sum_{k=1}^{n-2} v_k} \mathrm{d}v_{n-1}$$
$$\times \left[\prod_{j=1}^{n-1} \frac{v_j^{2/\alpha}}{v_j + \frac{\rho}{1-n\rho}} \right] \frac{(1 - v_1 - \cdots - v_{n-1})^{2/\alpha}}{(1 - v_1 - \cdots - v_{n-1}) + \frac{\rho}{1-n\rho}},$$
$$n = 1, 2, \ldots, \lceil \gamma^{-1} \rceil = \lceil \rho^{-1} \rceil - 1. \tag{4.75}$$

Note that $T_n(\rho,\alpha)$ for $n \geq 2$ is an $(n-1)$-dimensional integral. The first two terms in (4.74) can be evaluated in closed form (see Problem 4.16(1)):

$$T_n(\rho,\alpha) = \begin{cases} 1/(1+\gamma), & n=1, \\ \dfrac{8\sqrt{\pi}\,\Gamma\left(\frac{2}{\alpha}\right)}{4^{2/\alpha}\alpha^2\,\Gamma\left(\frac{3}{2}+\frac{2}{\alpha}\right)}\left(\dfrac{1-\gamma}{1+\gamma}\right)^2 {}_2F_1\left(\dfrac{1}{2},1;\dfrac{3}{2}+\dfrac{2}{\alpha};\left(\dfrac{1-\gamma}{1+\gamma}\right)^2\right), & n=2. \end{cases} \tag{4.76}$$

Note that we have expressed $T_1(\rho,\alpha)$ and $T_2(\rho,\alpha)$ directly in terms of γ instead of $\rho = \gamma/(1+\gamma)$ for convenience. In particular, when $\alpha = 4$ and $n = 2$, we have (see Problem 4.16(2))

$$T_2(\rho,4) = \frac{\pi}{2}\frac{(1-\sqrt{\gamma})^2}{(1+\gamma)}, \quad 0 < \gamma < 1. \tag{4.77}$$

Note that with two terms in (4.74), we already have the exact CCDF of the SIR for all γ such that $\lceil \gamma^{-1} \rceil \leq 2$, i.e. for all $\gamma \geq \frac{1}{2} = -3$ dB. For coverage in cellular networks, a common criterion is the requirement that the SINR should exceed $-6\,\text{dB} = \frac{1}{4}$ in at least 95% of locations in a cell (NTT DoCoMo, 2007). Assuming that thermal noise is negligible, this corresponds to the requirement that $\mathbb{P}\{\Gamma > \frac{1}{4}\} \geq 0.95$. Taking $\gamma \geq \frac{1}{4}$ in (4.74) yields $\lceil \gamma^{-1} \rceil = 4$ terms, of which the truncation to the first two terms is known from (4.76) to be

$$\alpha = 4,\ N_0 = 0 \Rightarrow \mathbb{P}\{\Gamma > \gamma\} = \frac{2}{\pi}\frac{1}{\sqrt{\gamma}} - \frac{1}{\pi}\left(\frac{1}{\sqrt{\gamma}} - 1\right)^2 1_{(0,1)}(\gamma),$$
$$\text{exact for } \gamma \geq \frac{1}{2}. \tag{4.78}$$

The third and fourth terms are, respectively, a two-dimensional and three-dimensional integral, which can be evaluated numerically from (4.75).

Important special case: $N_0 = 0$ and $\gamma \geq 1$

For $\gamma \geq 1$, $\lceil \gamma^{-1} \rceil = 1$, so only the $n = 1$ term survives in (4.72). Further, if $N_0 = 0$, we obtain the CCDF of the SIR to be (see Problem 4.17)

$$N_0 = 0 \Rightarrow \mathbb{P}\{\Gamma > \gamma\} = \frac{\operatorname{sinc}(2/\alpha)}{\gamma^{2/\alpha}}, \quad \gamma \geq 1. \tag{4.79}$$

This simple expression was first derived (for a multi-tier deployment) using a different argument by Dhillon *et al.* (2012). It was the first known expression for the distribution of the SIR at an arbitrary user in a cellular deployment under the max-SINR BS association rule. It is an important and tractable case, and has a special relationship to the max-SINR association rule in the sense that, if $\gamma > 1$, there is at most one BS whose received SINR at the user exceeds γ (see Problem 4.18). A generalization by Dhillon *et al.* (2012, Lem. 1) shows that at most n BSs exist such that the SINRs received at the user when receiving from them exceeds $1/n$, which matches what we know from (4.69).

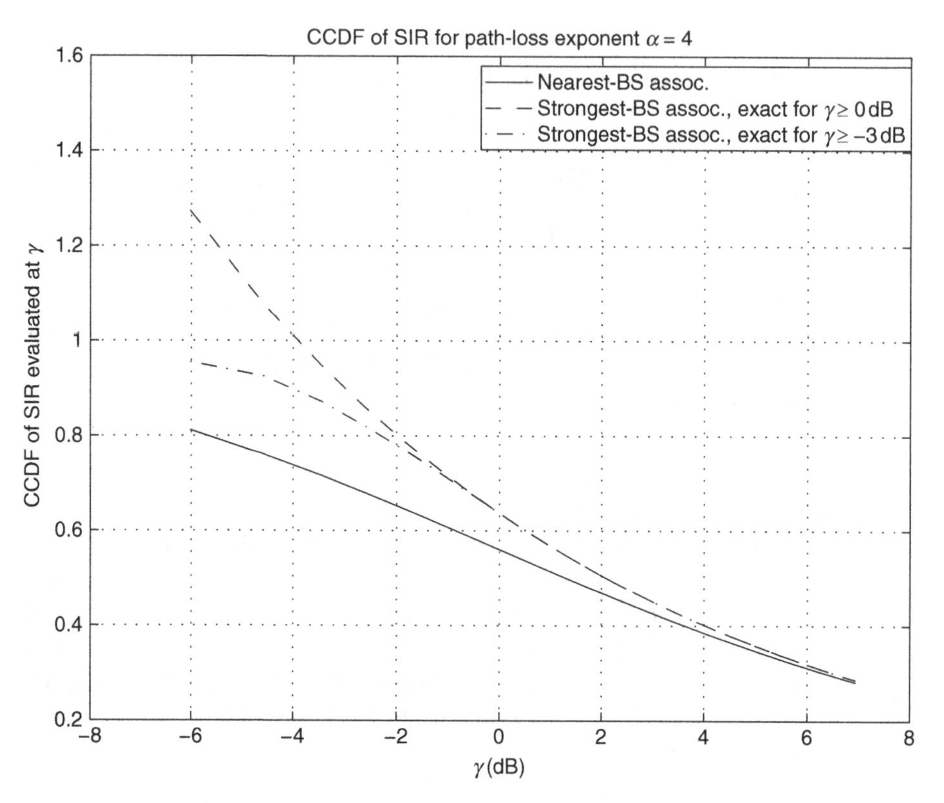

Figure 4.1 Plots of the CCDF of the SIR for nearest-BS and strongest-BS association rules for path-loss exponent $\alpha = 4$. Note that the CCDF for the nearest-BS association case is exact over all arguments γ, but holds only for i.i.d. Rayleigh fading. On the other hand, the CCDFs for the strongest-BS association case hold for arbitrary i.i.d. fading, but are exact over only the ranges of γ specified in the figure. The dashed-line curve is only exact for $\gamma \geq 0$ dB, and becomes more inaccurate as γ decreases, until it ceases to be usable for $\gamma < -4$ dB, when the value of the right side of (4.79) exceeds unity.

In Figure 4.1, we plot the CCDF of the SIR with path-loss exponent $\alpha = 4$ for the nearest-BS association rule given by (4.37) (which is exact, but holds only for i.i.d. Rayleigh fading), and the strongest-BS association rule given by (4.78) (which holds for arbitrary i.i.d. fading, but is only exact for arguments exceeding -3 dB). For comparison, we also plot the even simpler CCDF expression (4.79) for the strongest-BS association rule, which corresponds to the first term in (4.78), and is only exact for arguments exceeding 0 dB.

Joint CCDF of SINRs from *k* strongest BSs

In this section, we demonstrate the power of the canonical probability formulation of the joint CCDF problem by outlining the derivation of the joint CCDF of the SINR at a user when receiving from the strongest, second-strongest, ..., kth-strongest BSs.[2] We

[2] This is an advanced section and may be skipped on a first reading. It is best read after reading Section 5.3.1.

do not provide this derivation for the nearest-BS association rule, deferring that result to Section 5.3.1 for the more general case of multiple tiers (see (5.83)). On the other hand, the more general case of multiple tiers with the strongest-BS association rule for candidate serving BS selection in the tiers is not discussed in this book, and may be treated as an exercise for the reader.

We return to our model of a single tier of BSs, whose locations are modeled as the points of a homogeneous PPP Φ with density λ and i.i.d. Rayleigh fading on all links to the user from all BSs. Let us label the points of Φ by $\Phi = \{b_1, b_2, \dots\}$, and let us denote the kth strongest BS by B_k, $k = 1, 2, \dots$. Thus the serving BS, being the strongest BS, is $B \equiv B_1$. We may now write the joint CCDF of the SINRs from the first k strongest BSs for any k to be

$$
\begin{aligned}
&\mathbb{P}\left\{\frac{Y_{B_l}}{N_0 + \sum_{b \in \Phi \setminus \{B_l\}} Y_b} > \gamma_l,\ l = 1, \dots, k\right\} \\
&= \sum_{\substack{b_{i_1}, \dots, b_{i_k} \in \Phi: \\ (\forall l \neq m) b_{i_l} \neq b_{i_m}}} \mathbb{P}\left\{B_l = b_{i_l},\ \frac{Y_{b_{i_l}}}{N_0 + \sum_{b \in \Phi \setminus \{b_{i_l}\}} Y_b} > \gamma_l,\ l = 1, \dots, k\right\} \\
&= \sum_{\substack{b_{i_1}, \dots, b_{i_k} \in \Phi: \\ (\forall l \neq m) b_{i_l} \neq b_{i_m}}} \mathbb{P}\left(\left\{B_l = b_{i_l},\ \frac{Y_{b_{i_l}}}{N_0 + \sum_{b \in \Phi \setminus \{b_{i_l}\}} Y_b} > \gamma_l,\ l = 1, \dots, k\right\} \cap \right. \\
&\qquad \left. \bigcap_{b \in \Phi \setminus \{b_{i_1}, \dots, b_{i_k}\}} \{Y_b \leq Y_{b_{i_k}}\}\right) \\
&= \sum_{\substack{b_{i_1}, \dots, b_{i_k} \in \Phi: \\ (\forall l \neq m) b_{i_l} \neq b_{i_m}}} \mathbb{P}\left\{B_l = b_{i_l},\ \frac{Y_{b_{i_l}}}{N_0 + \sum_{b \in \Phi \setminus \{b_{i_l}\}} Y_b} > \gamma_l,\ l = 1, \dots, k\right\} \\
&\quad - \sum_{n=1}^{\infty} \frac{(-1)^{n-1}}{n!} \sum_{\substack{b_{i_1}, \dots, b_{i_{k+n}} \in \Phi: \\ (\forall l \neq m)\, b_{i_l} \neq b_{i_m}}} \mathbb{P}\left\{B_l = b_{i_l},\ \frac{Y_{b_{i_l}}}{N_0 + \sum_{b \in \Phi \setminus \{b_{i_l}\}} Y_b} > \gamma_l,\right. \\
&\qquad \left. l = 1, \dots, k,\ Y_{b_{i_{k+j}}} > Y_{b_{i_k}},\ j = 1, \dots, n\right\} \\
&= \sum_{\substack{b_{i_1}, \dots, b_{i_k} \in \Phi: \\ (\forall l \neq m) b_{i_l} \neq b_{i_m}}} \mathbb{P}\left\{\frac{Y_{b_{i_l}}}{N_0 + \sum_{b \in \Phi \setminus \{b_{i_l}\}} Y_b} > \gamma_l',\ l = 1, \dots, k\right\} \\
&\quad - \sum_{n=1}^{\infty} \frac{(-1)^{n-1}}{n!} \sum_{\substack{b_{i_1}, \dots, b_{i_{k+n}} \in \Phi: \\ (\forall l \neq m)\, b_{i_l} \neq b_{i_m}}} \mathbb{P}\left\{\frac{Y_{b_{i_l}}}{N_0 + \sum_{b \in \Phi \setminus \{b_{i_l}\}} Y_b} > \gamma_l',\right. \\
&\qquad \left. l = 1, \dots, k,\ Y_{b_{i_{k+j}}} > Y_{b_{i_k}},\ j = 1, \dots, n\right\},
\end{aligned}
$$

where we have used the fact that $B_l = b_{i_l} \Leftrightarrow Y_{b_{i_l}} > Y_{b_{i_{l+1}}}$, $l = 1, \ldots, k-1$, and

$$\frac{Y_{b_{i_l}}}{N_0 + \sum_{b \in \Phi \setminus \{b_{i_l}\}} Y_b} > \gamma_l \Leftrightarrow \frac{Y_{b_{i_l}}}{N_0 + \sum_{l=1}^{k} Y_{b_{i_l}} + \sum_{b \in \Phi \setminus \{b_{i_1}, \ldots, b_{i_k}\}} Y_b} > \rho_l, \; l = 1, \ldots, k,$$

together with the definitions

$$\rho_l = \frac{\gamma_l}{1+\gamma_l}, \quad \gamma_l' = \frac{\rho_l'}{1-\rho_l'}, \quad \rho_l' = \begin{cases} \max\{\rho_l, \rho_{l+1}\}, & l = 1, \ldots, k-1, \\ \rho_k, & l = k. \end{cases}$$

Define $\boldsymbol{\rho}' = [\rho_1', \ldots, \rho_k']^\top$. Then we can write

$$\sum_{\substack{b_{i_1}, \ldots, b_{i_k} \in \Phi: \\ (\forall l \neq m) b_{i_l} \neq b_{i_m}}} \mathbb{P}\left\{ \frac{Y_{b_{i_l}}}{N_0 + \sum_{b \in \Phi \setminus \{b_{i_l}\}} Y_b} > \gamma_l', \; l = 1, \ldots, k \right\}$$
$$= \sum_{\substack{b_{i_1}, \ldots, b_{i_k} \in \Phi: \\ (\forall l \neq m) b_{i_l} \neq b_{i_m}}} \mathbb{P}\left\{ \left(\mathbf{I}_k - \boldsymbol{\rho}' \mathbf{1}_k^\top\right) [Y_{b_{i_1}}, \ldots, Y_{b_{i_k}}]^\top > \left(\sum_{b \in \Phi \setminus \{b_{i_1}, \ldots, b_{i_k}\}} Y_b \right) \boldsymbol{\rho}' \right\} \tag{4.80}$$

and

$$\sum_{\substack{b_{i_1}, \ldots, b_{i_{k+n}} \in \Phi: \\ (\forall l \neq m) \, b_{i_l} \neq b_{i_m}}} \mathbb{P}\left\{ \frac{Y_{b_{i_l}}}{N_0 + \sum_{b \in \Phi \setminus \{b_{i_l}\}} Y_b} > \gamma_l', \; l = 1, \ldots, k, \; Y_{b_{i_{k+j}}} > Y_{b_{i_k}}, \; j = 1, \ldots, n \right\}$$
$$= \sum_{\substack{b_{i_1}, \ldots, b_{i_{k+n}} \in \Phi: \\ (\forall l \neq m) \, b_{i_l} \neq b_{i_m}}} \mathbb{P}\left\{ \mathbf{A}_{k+n} [Y_{b_{i_1}}, \ldots, Y_{b_{i_{k+n}}}]^\top > \left(\sum_{b \in \Phi \setminus \{b_{i_1}, \ldots, b_{i_{k+n}}\}} Y_b \right) [\boldsymbol{\rho}'^\top, \mathbf{0}_n^\top]^\top \right\}, \tag{4.81}$$

where (see (5.88))

$$\mathbf{A}_{k+n} = \begin{bmatrix} \mathbf{I}_k - \boldsymbol{\rho}' \mathbf{1}_k^\top & -\boldsymbol{\rho}' \mathbf{1}_n^\top \\ -\mathbf{1}_n^\top \left(\boldsymbol{e}_k^{(k)}\right)^\top & \mathbf{I}_n \end{bmatrix}$$

can be shown to be a $(k+n) \times (k+n)$ Z-matrix satisfying (2.25), whose determinant and inverse can both be calculated analytically.

The summations in (4.80) and (4.81) can be evaluated using Theorem 4.15 following the same steps as in the derivation of (4.66) from (4.64). We do not provide the details here, noting that they will become clearer in the context of the derivation of the joint CCDF of the SINR in multi-tier deployments provided in Chapter 5 (see Section 5.3.1).

Problems

4.13 Derive (4.51) from (4.50) as follows.

(1) Use (4.52) to write the integral on the right side of (4.50) as follows:

$$\int_0^{y^*} [1-\exp(-sy^*)]\tilde{\lambda}(y)\,\mathrm{d}y = \pi\nu P^{2/\alpha}\,\mathbb{E}[H^{2/\alpha}]\,\frac{2}{\alpha}\int_0^{y^*}\frac{1-\exp(-sy^*)}{y^{1+2/\alpha}}\,\mathrm{d}y\,. \tag{4.82}$$

(2) Use (4.41) to write

$$\frac{2}{\alpha}\int_0^{y^*}\frac{1-\exp(-sy^*)}{y^{1+2/\alpha}}\,\mathrm{d}y = s^{2/\alpha}\left[\gamma\left(1-\frac{2}{\alpha}, sy^*\right) - \frac{1-\exp(-sy^*)}{(sy^*)^{2/\alpha}}\right]. \tag{4.83}$$

(3) Substitute (4.82) and (4.83) into (4.50) to get (4.51).

4.14 Verify (4.69). *Hint*: Verify separately for $\gamma = 0$, $\gamma > 1$, $\gamma = 1$, and $0 < \gamma < 1$.

4.15 Derive (4.74) and (4.75) as follows.

(1) In (4.72) with $N_0 = 0$, make the change of variables

$$t = \left(\sum_{k=1}^{n} u_k^{\alpha/2}\right)^{2/\alpha}, \quad v_k = \left(\frac{u_k}{t}\right)^{\alpha/2}, \quad k = 1,\ldots,n-1.$$

Prove that the Jacobian of this transformation is given by

$$(2/\alpha)^{n-1} t^{n-1} v_1^{2/\alpha-1}\cdots v_{n-1}^{2/\alpha-1}(1-v_1-\cdots-v_{n-1})^{2/\alpha-1}.$$

Hint: Extract the multiplying factors from each row of the matrix of partial derivatives, and the multiplying factor from each column but the first, then use the fact that the determinant of a matrix is unchanged if the last row is replaced by the sum of all the rows.

(2) Verify that the region of integration over $(u_1,\ldots,u_n)$ in (4.72) becomes $(0,\infty)\times\mathcal{S}_{n-1}$ over $(t, v_1,\ldots,v_{n-1})$, where $\mathcal{S}_{n-1}$ is the unit simplex in $\mathbb{R}^{n-1}$, given by

$$\mathcal{S}_{n-1} = \{(v_1,\ldots,v_{n-1}) \in \mathbb{R}_+^{n-1} : v_1+\cdots+v_{n-1} \le 1\}.$$

(3) Rewrite the integral in (4.72) with the transformation of variables in (1) over the region in (2). Then perform the integration with respect to t, and show that (4.72) can then be written as (4.74), where $T_n(\rho,\alpha)$ is given by (4.75).

4.16 (1) Prove (4.76). *Hint*: Prove directly from (4.72) for $n = 1$. For $n = 2$, show that (4.75) yields

$$T_2(\rho,\alpha) = \frac{2}{\alpha}\int_0^1 \frac{v_1^{2/\alpha}(1-v_1)^{2/\alpha}}{\left(v_1+\frac{\gamma}{1-\gamma}\right)\left(1-v_1+\frac{\gamma}{1-\gamma}\right)}\,\mathrm{d}v_1,$$

then evaluate this integral by successively applying Gradshteyn & Ryzhik (2000, 3.197.3, 9.134.1, 8.335.1).

(2) Apply Gradshteyn & Ryzhik (2000, 9.121.24) to (4.76) with $(1-\gamma)^2/(1+\gamma)^2 = 4z(1-z) = 1-(1-2z)^2$ for $z = (1-\sqrt{\gamma})^2/[2(1+\gamma)]$ to prove (4.77).

4.17 Prove (4.79). *Hint*: Substitute $T_1(\rho, \alpha)$ from (4.76) into (4.74).

4.18 Prove that there can be at most one BS in a (possibly multi-tier) deployment such that the received SINR at the user exceeds 1. *Hint*: Suppose there are two BSs such that the received SINR from each of them exceeds 1. Let the received powers from these two BSs be U_1 and U_2, say, and the total received power from all *other* BSs in the deployment, plus thermal noise, be V. Then, if the received SINR from both BSs exceeds 1, we must have $U_1/(U_2 + V) > 1$, i.e. $U_1 > U_2 + V > U_2$, and similarly $U_2 > U_1 + V > U_1$, which is impossible.

5 SINR analysis for multiple tiers with fixed powers

5.1 Introduction

The theme of this book is the analysis of the distribution of SINR in a multi-tier HCN. We introduced the model and the basic notation in Section 2.1.2. However, before we could begin to calculate the joint CCDF (2.6), we first needed to study the structure of the canonical probability problem, then understand how the PPP model for BS locations in the tiers made the canonical probability mathematically tractable. Finally, we applied those skills and insights to the derivation of the CCDF of the SINR for a single-tier deployment. We are now at the point where we can pick up the calculation of the joint CCDF (2.6) where we left it in Section 2.2. We shall use the notation previously introduced in Section 2.1.2, with the additional assumption that Φ_i, the point pattern describing the locations of the BSs belonging to any tier i, is a homogeneous PPP with density λ_i.

A crucial difference from the single-tier deployment scenario analyzed in Chapter 4 is the classification of tiers as either *open*, i.e. *accessible* to the user, or *closed*, i.e. not accessible to the user. Recall that, in the notation of Section 2.1.2, in a deployment with n_{tier} tiers of BSs, labeled $1, \ldots, n_{\text{tier}}$, the tiers that the user is allowed to access are labeled $1, \ldots, n_{\text{open}}$, where $n_{\text{open}} \leq n_{\text{tier}}$.

All BSs in a given tier i are assumed to be identical in terms of their capabilities, and to transmit with the same power P_i^{tx}, which need not be the same across tiers. Further, the heights of the BSs in different tiers may be different, thereby leading to different slope-intercept values for the path-loss models (1.1) for the links from BSs in the different tiers to the user.

Consider an arbitrarily located user in the network. Recall that there is a *candidate serving BS*, denoted $B_i \in \Phi_i$ for the user from each open tier i, $i = 1, \ldots, n_{\text{open}}$. The (actual) serving BS for the user is chosen from among these candidate serving BSs in some way, which we shall study in a later section. We use the notation of (2.1), except that the fade attenuation on the link from B_i to the user is denoted $H_{i,*}$ for brevity instead of H_{B_i}. Note that $H_{1,*}, \ldots, H_{n_{\text{open}},*}$ are independent, though not i.i.d. in general. We also assume that, for each open tier $i = 1, \ldots, n_{\text{open}}$, the fade attenuations H_b on the links to the user from the BSs $b \in \Phi_i \setminus \{B_i\}$ of tier i are i.i.d., independent of $H_{i,*}$, and independent of (but in general not distributed identically to) the fade attenuations on the links to the user from the BSs of any other tier $j \neq i$.

Similarly, for each closed tier $j = n_{\text{open}} + 1, \ldots, n_{\text{tier}}$, the fade attenuations H_b on the links to the user from the BSs $b \in \Phi_j$ of tier j are i.i.d. and independent of (but in general not distributed identically to) the fade attenuations on the links to the user from the BSs of any other tier $k \neq j$. Let us also define H_i to be a random variable with the same distribution as the i.i.d. fade attenuations on the links to the user from the non-candidate BSs $b \in \Phi_i \setminus \{B_i\}$ in tier i, $i = 1, \ldots, n_{\text{open}}$, and similarly let H_j be a random variable with the same distribution as the i.i.d. fade attenuations on the links to the user from any of the BSs in tier j, $j = n_{\text{open}} + 1, \ldots, n_{\text{tier}}$.

Throughout this chapter, we focus on the calculation of the joint CCDF (2.6) when the tiers are $i_1 = 1$, $i_2 = 2$, $\ldots$, $i_k = k$, for some $k \leq n_{\text{open}}$. For arbitrary distinct $i_1, \ldots, i_k$ such that $\{i_1, \ldots, i_k\} \subseteq \{1, \ldots, n_{\text{open}}\}$, we may, without loss of generality, *relabel* the open tiers such that $i_1 = 1$, $\ldots$, $i_k = k$, the remaining open tiers $\{1, \ldots, n_{\text{open}}\} \setminus \{i_1, \ldots, i_k\}$ in the original labeling now being labeled $k+1, \ldots, n_{\text{open}}$, respectively, and keep the closed tier labeling unchanged. Then the joint CCDF (2.6) under the original labeling of the tiers becomes the joint CCDF of $\Gamma_1, \ldots, \Gamma_k$ under the new labeling, evaluated at the arguments given by the same new labeling scheme applied to the original arguments $\gamma_{i_1}, \ldots, \gamma_{i_k}$ in (2.6).

Hence in what follows we focus on calculating the joint CCDF:

$$\mathbb{P}\{\boldsymbol{\Gamma} > \boldsymbol{\gamma}\}, \quad \boldsymbol{\gamma} > \mathbf{0}_k, \quad \boldsymbol{\Gamma} = [\Gamma_1, \ldots, \Gamma_k]^\top, \quad k \leq n_{\text{open}}.$$

5.2 Joint CCDF of SINR from candidate serving BSs

As in Chapter 4, we consider two cases, namely where the candidate serving BS in each tier is either the one that is nearest to the user (assumed located at the origin), or the one that is received most strongly at the user. We assume that the same candidate serving BS selection criterion (i.e. either nearest BS or most strongly received BS) is applied to each open tier. Note that, for a single-tier deployment, the "candidate" serving BS is also the actual serving BS, so the candidate serving BS selection criterion also yields the BS association rule. However, in a multi-tier deployment, neither of the two candidate serving BS selection criteria above is an *association rule*. The association rule in a multi-tier HCN is given by the overall serving BS selection criterion, which is studied in a later section, and, as we shall see, is considerably richer than the association rules we studied for the single-tier deployments in Chapter 4.

5.2.1 Candidate serving BS in each tier is the one nearest to the user

Conditional SINR distribution

For $i = 1, \ldots, k$ ($k \leq n_{\text{open}}$), let us denote by R_i the distance of the candidate serving (i.e. nearest) BS in tier i, denoted B_i, from the user, and let $\boldsymbol{R} = [R_1, \ldots, R_k]^\top$. From Section 2.3.1, we see that, conditioned on $\boldsymbol{R} = \boldsymbol{r}$, the joint CCDF is given by

$$\mathbb{P}\{\boldsymbol{\Gamma} > \boldsymbol{\gamma} \mid \boldsymbol{R} = \boldsymbol{r}\} = \mathbb{P}\{\mathbf{A}\boldsymbol{X} > W\boldsymbol{\rho} \mid \boldsymbol{R} = \boldsymbol{r}\}, \tag{5.1}$$

where

$$\boldsymbol{X} = [X_1, \ldots, X_k]^\top, \quad X_i = \frac{P_i H_{i,*}}{r_i^{\alpha_i}}, \quad P_i = K_i P_i^{\text{tx}}, \quad i = 1, \ldots, k,$$

and from (2.10), (2.21), and (2.11) we have

$$\mathbf{A} = \mathbf{I}_k - \boldsymbol{\rho} \mathbf{1}_k^\top, \quad \boldsymbol{\rho} = \left[\frac{\gamma_1}{1+\gamma_1}, \ldots, \frac{\gamma_k}{1+\gamma_k} \right]^\top,$$

and

$$W = \sum_{i=1}^{k} V_i + \sum_{j=k+1}^{n_{\text{tier}}} W_j + N_0, \tag{5.2}$$

where

$$V_i = \sum_{b \in \Phi_i \setminus \{B_i\}} \frac{P_i H_b}{R_b^{\alpha_i}}, \; i = 1, \ldots, n_{\text{open}}, \quad W_j = \sum_{b \in \Phi_j} \frac{P_j H_b}{R_b^{\alpha_j}}, \; j = 1, \ldots, n_{\text{tier}}.$$

We can show (see Problem 5.1) that if $H_{1,*}, \ldots, H_{k,*}$ i.i.d. Exp(1), the right side of (5.1) may be evaluated as follows (Mukherjee, 2012a, Thm. 3):

$$\mathbb{P}\{\boldsymbol{\Gamma} > \boldsymbol{\gamma} \mid \boldsymbol{R} = \boldsymbol{r}\}$$
$$= \begin{cases} 0, & \sum_{i=1}^{k} \gamma_i/(1+\gamma_i) \geq 1, \\ \dfrac{1}{\left(1 - \sum_{i=1}^{k} \dfrac{\gamma_i}{1+\gamma_i}\right)} \dfrac{\mathcal{L}_{W \mid \boldsymbol{R}=\boldsymbol{r}}(s)}{\prod_{i=1}^{k} \left(1 + \dfrac{s}{c_i}\right)} \Bigg|_{s = \frac{\boldsymbol{c}^\top \boldsymbol{\rho}}{1 - \mathbf{1}_k^\top \boldsymbol{\rho}}}, & \sum_{i=1}^{k} \gamma_i/(1+\gamma_i) < 1, \end{cases} \tag{5.3}$$

where

$$c_i = \frac{r_i^{\alpha_i}}{P_i}, \; i = 1, \ldots, k, \quad \boldsymbol{c} = [c_1, \ldots, c_k]^\top. \tag{5.4}$$

From the independence of the fading across the tiers, (5.2) yields

$$\begin{aligned} \mathcal{L}_{W \mid \boldsymbol{R}=\boldsymbol{r}}(s) &= \mathrm{e}^{-sN_0} \prod_{i=1}^{k} \mathcal{L}_{V_i \mid R_i = r_i}(s) \prod_{j=k+1}^{n_{\text{tier}}} \mathcal{L}_{W_j}(s) \\ &= \mathrm{e}^{-sN_0} \prod_{i=1}^{k} \exp\left[-\pi \lambda_i (P_i s)^{2/\alpha_i} G_{i,\alpha_i}\left(\frac{r_i^2}{P_i^{2/\alpha_i} s^{2/\alpha_i}} \right) \right] \\ &\times \prod_{j=k+1}^{n_{\text{tier}}} \exp\left\{ -\pi \lambda_j (P_j s)^{2/\alpha_j} G_{j,\alpha_j}(0) \right\}, \end{aligned} \tag{5.5}$$

where we have used (4.24) and (4.11), and

$$G_{i,\alpha}(z) = \mathbb{E}\left[\gamma\left(1 - \frac{2}{\alpha}, \frac{H_i}{z^{\alpha/2}}\right) H_i^{2/\alpha} \right] - z\left[1 - \mathcal{L}_{H_i}\left(\frac{1}{z^{\alpha/2}}\right)\right],$$
$$z \geq 0, \quad i = 1, \ldots, n_{\text{open}}. \tag{5.6}$$

We therefore have the following result.

THEOREM 5.1 *If the candidate serving BSs from the open tiers are those nearest to the user, and if the fade attenuations on the links from the candidate serving BSs in the open tiers to the user are i.i.d. exponentially distributed with unit mean, then, conditioned on the distances to these candidate serving BSs, for any* $k \le n_{\text{open}}$ *we have*

$$\mathbb{P}\{\Gamma_1 > \gamma_1, \ldots, \Gamma_k > \gamma_k \mid R_1 = r_1, \ldots, R_k = r_k\}$$
$$= \begin{cases} 0, & \sum_{i=1}^{k} \gamma_i/(1+\gamma_i) \ge 1, \\ \dfrac{F_{\boldsymbol{r}}\left(\dfrac{1}{1 - \sum_{j=1}^{k} \gamma_j/(1+\gamma_j)} \sum_{i=1}^{k} \dfrac{r_i^{\alpha_i} \gamma_i}{P_i(1+\gamma_i)} \right)}{1 - \sum_{j=1}^{k} \gamma_j/(1+\gamma_j)}, & \sum_{i=1}^{k} \gamma_i/(1+\gamma_i) < 1, \end{cases} \tag{5.7}$$

where

$$F_{\boldsymbol{r}}(s) = \left[\prod_{i=1}^{k} \left(1 + \frac{P_i s}{r_i^{\alpha_i}} \right) \right]^{-1} \exp\left\{ -sN_0 - \pi \sum_{i=1}^{k} \lambda_i (P_i s)^{2/\alpha_i} G_{i,\alpha_i}\left(\frac{r_i^2}{P_i^{2/\alpha_i} s^{2/\alpha_i}} \right) \right.$$
$$\left. - \pi \sum_{j=k+1}^{n_{\text{tier}}} \lambda_j (P_j s)^{2/\alpha_j} G_{j,\alpha_j}(0) \right\}, \quad s > 0,$$

and the functions $G_{i,\alpha}(\cdot)$, $i = 1, \ldots, n_{\text{open}}$, *are as defined in* (5.6).

Theorem 5.1 was first proved in Mukherjee (2012a). When $H_{1,*}, \ldots, H_{n_{\text{open}},*}$ are independent but have arbitrary distributions, we can use Theorem 2.5 together with the hyper-Erlang approximations to the corresponding distributions of $X_i = P_i H_{i,*}/r_i^{\alpha_i}$, $i = 1, \ldots, n_{\text{open}}$, but it is more useful to recast the joint CCDF expressions directly in terms of the distributions of $H_{1,*}, \ldots, H_{n_{\text{open}},*}$.

THEOREM 5.2 *If the candidate serving BSs from the open tiers are those nearest to the user, and if the fade attenuation on the link from the candidate serving BS in tier i has PDF* $f_{H_{i,*}}(\cdot)$, $i = 1, \ldots, n_{\text{open}}$, *then, conditioned on the distances to these candidate serving BSs, for any* $k \le n_{\text{open}}$ *we have*

$$\mathbb{P}\{\Gamma_1 > \gamma_1, \ldots, \Gamma_k > \gamma_k \mid R_1 = r_1, \ldots, R_k = r_k\}$$
$$= \begin{cases} 0, & \sum_{i=1}^{k} \gamma_i/(1+\gamma_i) \ge 1, \\ \displaystyle\sum_{l_1=1}^{\infty} \cdots \sum_{l_k=1}^{\infty} \left\{ \prod_{n=1}^{k} \frac{(-1)^{m_{n,l_n}} \beta_{n,l_n} (c_{n,l_n})^{m_{n,l_n}+1}}{m_{n,l_n}!} \right\} & \\ \times \left(1 - \displaystyle\sum_{i=1}^{k} \frac{\gamma_i}{1+\gamma_i} \right)^{-1} \dfrac{\partial^{m_{1,l_1}+\cdots+m_{k,l_k}}}{\partial (c_{1,l_1})^{m_{1,l_1}} \cdots \partial (c_{k,l_k})^{m_{k,l_k}}} & \\ \dfrac{\mathcal{L}_{W \mid \boldsymbol{R}=\boldsymbol{r}}\left(\dfrac{1}{1 - \sum_{j=1}^{k} \gamma_j/(1+\gamma_j)} \displaystyle\sum_{i=1}^{k} \frac{c_{i,l_i} r_i^{\alpha_i} \gamma_i}{P_i(1+\gamma_i)} \right)}{\prod_{i=1}^{k} \left(c_{i,l_i} + \frac{P_i/r_i^{\alpha_i}}{1 - \sum_{j=1}^{k} \gamma_j/(1+\gamma_j)} \sum_{n=1}^{k} \frac{c_{n,l_n} r_n^{\alpha_n} \gamma_n}{P_n(1+\gamma_n)} \right)}, & \sum_{i=1}^{k} \gamma_i/(1+\gamma_i) < 1, \end{cases}$$

(5.8)

where each PDF $f_{H_{i,*}}(\cdot)$, $i = 1, \dots, n_{\text{open}}$, *is expressed as an infinite mixture of Erlang PDFs in the form*

$$f_{H_{i,*}}(x) = \sum_{l=1}^{\infty} \beta_{i,l} \frac{c_{i,l}^{m_{i,l}+1}}{m_{i,l}!} x^{m_{i,l}} \exp(-c_{i,l}x),\ x \geq 0, \quad m_{i,l} \in \{0, 1, \dots\},\ \sum_{l=1}^{\infty} \beta_{i,l} = 1, \tag{5.9}$$

and $\mathcal{L}_{W\,|\,\boldsymbol{R}=\boldsymbol{r}}(s)$ *is given by* (5.5).

Proof Let us define

$$\mu_{i,*} = \frac{P_i}{r_i^{\alpha_i}}, \quad i = 1, \dots, k, \quad \boldsymbol{\mu}_* = \left[\mu_{1,*}, \dots, \mu_{k,*}\right]^\top. \tag{5.10}$$

Then in (5.1) we have

$$\boldsymbol{X} = \mathbf{M}_*\boldsymbol{H}_*, \quad \mathbf{M}_* = \begin{bmatrix} \mu_{1,*} & 0 & \cdots & 0 \\ 0 & \mu_{2,*} & \cdots & 0 \\ \vdots & \vdots & \ddots & \vdots \\ 0 & \cdots & 0 & \mu_{k,*} \end{bmatrix} = \operatorname{diag}(\boldsymbol{\mu}_*). \tag{5.11}$$

Thus the inequality in (5.1) can be rewritten as

$$\mathbf{A}\mathbf{M}_*\boldsymbol{H}_* > W\boldsymbol{\rho} \Leftrightarrow \mathbf{A}_*\boldsymbol{H}_* > W\boldsymbol{\rho}_*,$$

where

$$\boldsymbol{\rho}_* = \mathbf{M}_*^{-1}\boldsymbol{\rho} = \left[\frac{\rho_1}{\mu_{1,*}}, \dots, \frac{\rho_k}{\mu_{k,*}}\right]^\top,$$
$$\mathbf{A}_* = \mathbf{M}_*^{-1}\mathbf{A}\mathbf{M}_* = \mathbf{M}_*^{-1}\left(\mathbf{I}_k - \boldsymbol{\rho}\mathbf{1}_k^\top\right)\mathbf{M}_* = \mathbf{I}_k - \boldsymbol{\rho}_*\boldsymbol{\mu}_*^\top.$$

It follows that (2.25) holds for the Z-matrix $\mathbf{A}_*$, where

$$\det \mathbf{A}_* = \det \mathbf{M}_*^{-1} \det \mathbf{A} \det \mathbf{M}_* = \det \mathbf{A} = 1 - \mathbf{1}_k^\top \boldsymbol{\rho}$$

and

$$\mathbf{A}_*^{-1} = \mathbf{M}_*^{-1}\mathbf{A}^{-1}\mathbf{M}_* = \mathbf{I}_k + \frac{1}{1 - \mathbf{1}_k^\top\boldsymbol{\rho}}\boldsymbol{\rho}_*\boldsymbol{\mu}_*^\top = \mathbf{I}_k + \frac{1}{1 - \sum_{i=1}^k \gamma_i/(1+\gamma_i)}\boldsymbol{\rho}_*\boldsymbol{\mu}_*^\top.$$

The theorem follows by applying Theorem 2.5 to the calculation of

$$\mathbb{P}\{\boldsymbol{\Gamma} > \boldsymbol{\gamma} \mid \boldsymbol{R} = \boldsymbol{r}\} = \mathbb{P}\{\mathbf{A}_*\boldsymbol{H}_* > W\boldsymbol{\rho}_* \mid \boldsymbol{R} = \boldsymbol{r}\}$$

and using the results that, for any $l_1, \dots, l_k$ and $\boldsymbol{c} = \left[c_{1,l_1}, \dots, c_{k,l_k}\right]^\top$,

$$\begin{aligned} \boldsymbol{c}^\top\mathbf{A}_*^{-1}\boldsymbol{e}_i^{(k)} &= c_{i,l_i} + \mu_{i,*}\frac{\boldsymbol{c}^\top\boldsymbol{\rho}_*}{1 - \sum_{j=1}^k \gamma_j/(1+\gamma_j)} \\ &= c_{i,l_i} + \frac{P_i/r_i^{\alpha_i}}{1 - \sum_{j=1}^k \gamma_j/(1+\gamma_j)} \sum_{n=1}^k \frac{c_{n,l_n} r_n^{\alpha_n} \gamma_n}{P_n(1+\gamma_n)}, \quad i = 1, \dots, k, \end{aligned}$$

and

$$c^\top \mathbf{A}_*^{-1} \boldsymbol{\rho}_* = \frac{1}{1 - \sum_{j=1}^{k} \gamma_j/(1+\gamma_j)} c^\top \boldsymbol{\rho}_* = \frac{1}{1 - \sum_{j=1}^{k} \gamma_j/(1+\gamma_j)} \sum_{i=1}^{k} \frac{c_{i,l_i} r_i^{\alpha_i} \gamma_i}{P_i(1+\gamma_i)}.$$

□

Remark 5.1 As for the single-tier deployment, in practice the infinite mixture in (5.9) for each $i = 1, \dots, n_{\text{open}}$ would be approximated by a finite mixture by truncating the infinite series after a finite number of terms, perhaps using one of the methods mentioned in Remark 2.4. This also turns the multiple-infinite series in (5.8) into a finite sum.

Unconditional SINR distribution

As for the single-tier deployment, the unconditional joint CCDF of the SINR at the user location from the candidate serving BSs in the open tiers is obtained from the conditional joint CCDF by integrating over the joint PDF of the distances from the user to these candidate serving BSs:

$$\mathbb{P}\{\boldsymbol{\Gamma} > \boldsymbol{\gamma}\} = \int_{\mathbb{R}_+^k} \mathbb{P}\{\boldsymbol{\Gamma} > \boldsymbol{\gamma} \mid \boldsymbol{R} = \boldsymbol{r}\} f_{\boldsymbol{R}}(\boldsymbol{r}) \mathrm{d}\boldsymbol{r}. \tag{5.12}$$

Further, the independence of the PPPs modeling the BS locations in the tiers means that the joint PDF of the distance from the user to the nearest BSs in the tiers is just the product of the marginal PDFs of the distance from the user to the nearest BS in each tier, which are given by (4.34). Thus we have the joint PDF of $\boldsymbol{R}$, the random vector of distances from the user to the nearest BSs in the open tiers, given by

$$f_{\boldsymbol{R}}(\boldsymbol{r}) = \prod_{i=1}^{k} 2\pi\lambda_i r_i \exp\left(-\pi\lambda_i r_i^2\right), \quad \boldsymbol{r} = [r_1, \dots, r_k]^\top \in \mathbb{R}_+^k. \tag{5.13}$$

Making the change of variables from $\boldsymbol{r}$ to $\boldsymbol{u} = [u_1, \dots, u_k]^\top$, where $u_i = \pi\lambda_i r_i^2$, $i = 1, \dots, k$, gives us

$$f_{\boldsymbol{R}}(\boldsymbol{r}) \mathrm{d}\boldsymbol{r} = \exp\left(-\mathbf{1}_k^\top \boldsymbol{u}\right) \mathrm{d}\boldsymbol{u}. \tag{5.14}$$

With this change of variables, we may rewrite (5.5) as follows:

$$\mathcal{L}_{W \mid \boldsymbol{R}=\boldsymbol{r}}(s) = \exp\Bigg\{-sN_0 - \sum_{i=1}^{k} u_i \left(\frac{P_i s}{r_i^{\alpha_i}}\right)^{2/\alpha_i} G_{i,\alpha_i}\left[\left(\frac{r_i^{\alpha_i}}{P_i s}\right)^{2/\alpha_i}\right] - s^{2/\alpha_j}\pi \sum_{j=k+1}^{n_{\text{tier}}} \left(\lambda_j^{\alpha_j/2} P_j\right)^{2/\alpha_j} \mathbb{E}\left[H_j^{2/\alpha_j}\right] \Gamma\left(1 - \frac{2}{\alpha_j}\right)\Bigg\}, \quad s > 0. \tag{5.15}$$

With the expansion of the PDFs $f_{H_{i,*}}(\cdot)$ in the form (5.9), $i = 1,\ldots,k$, and for any $l_1,\ldots,l_k$, the last term on the right of (5.8) is $\mathcal{L}_{W|\boldsymbol{R}=\boldsymbol{r}}(s)/\left[\prod_{i=1}^{k}\left(c_{i,l_i} + P_i s/r_i^{\alpha_i}\right)\right]$, where

$$\begin{aligned} s &= \frac{1}{1-\sum_{j=1}^{k}\gamma_j/(1+\gamma_j)}\sum_{n=1}^{k}\frac{c_{n,l_n} r_n^{\alpha_n}\gamma_n}{P_n(1+\gamma_n)} \\ &= \frac{1}{1-\sum_{j=1}^{k}\gamma_j/(1+\gamma_j)}\sum_{n=1}^{k}\frac{c_{n,l_n} u_n^{\alpha_n/2}\gamma_n}{\pi^{\alpha_n/2}\lambda_n^{\alpha_n/2}P_n(1+\gamma_n)}. \end{aligned} \tag{5.16}$$

From (5.16), we see that, for any $i = 1,\ldots,k$,

$$\frac{P_i s}{r_i^{\alpha_i}} = \left[\frac{(P_i s)^{2/\alpha_i}}{r_i^2}\right]^{\alpha_i/2} = \frac{\pi^{\alpha_i/2}\lambda_i^{\alpha_i/2}P_i/u_i^{\alpha_i/2}}{1-\sum_{j=1}^{k}\gamma_j/(1+\gamma_j)}\sum_{n=1}^{k}\frac{c_{n,l_n}}{\pi^{\alpha_n/2}}\frac{\gamma_n}{(1+\gamma_n)}\frac{u_n^{\alpha_n/2}}{\lambda_n^{\alpha_n/2}P_n}. \tag{5.17}$$

From (5.16), (5.17), and (5.15), we see that $\mathcal{L}_{W|\boldsymbol{R}=\boldsymbol{r}}(s)/\left[\prod_{i=1}^{k}\left(c_{i,l_i} + P_i s/r_i^{\alpha_i}\right)\right]$ can be expressed entirely as a function of $\boldsymbol{u}$ parameterized by the thermal noise power N_0 and $\lambda_i^{\alpha_i/2}K_iP_i^{\text{tx}}$, $i = 1,\ldots,n_{\text{tier}}$. From (5.8) and (5.12), we then have the following result.

THEOREM 5.3 *With the hypotheses of Theorem 5.2, the unconditional joint CCDF of the SINRs at the user from the candidate serving BSs in the tiers depends only upon the thermal noise power N_0 and the deployment parameters of the tiers in the forms $\lambda_i^{\alpha_i/2}K_iP_i^{\text{tx}}$, $i = 1,\ldots,n_{\text{tier}}$.*

Remark 5.2 As in the derivation of (4.39) for the single-tier deployment, if we are interested in the joint SINR distribution at a user located at a minimum distance of $r_{\min,i}$ from the candidate serving BS in tier i, $i = 1,\ldots,k$, then the lower limit of integration over r_i in (5.8) is $r_{\min,i}$, $i = 1,\ldots,k$. If, in addition, we also require the user to be located at a minimum distance of, say, $d_{\min,j}$ from all BSs (and thus the nearest BS) in the non-serving tiers $k+1,\ldots,n_{\text{tier}}$, then from (4.24) and (4.27) we see that (5.15) changes to

$$\begin{aligned} \mathcal{L}_{W|\boldsymbol{R}=\boldsymbol{r}}(s) = \exp\Bigg\{ &-sN_0 - \sum_{i=1}^{k} u_i\left(\frac{P_i s}{r_i^{\alpha_i}}\right)^{2/\alpha_i} G_{i,\alpha_i}\left[\left(\frac{r_i^{\alpha_i}}{P_i s}\right)^{2/\alpha_i}\right] \\ &- \pi\sum_{j=k+1}^{n_{\text{tier}}} s^{2/\alpha_j}\left(\lambda_j^{\alpha_j/2}P_j\right)^{2/\alpha_j} G_{j,\alpha_j}\left[\left(\frac{d_{\min,j}^{\alpha_j}}{P_j s}\right)^{2/\alpha_j}\right]\Bigg\}, \quad s > 0, \end{aligned} \tag{5.18}$$

which is then used in (5.8). It is important to note that (5.8) is the expression for the conditional CCDF. Thus the minimum distance requirement on the other tiers $k+1,\ldots,n_{\text{tier}}$ changes both the conditional and unconditional joint CCDFs of the SINRs from the candidate serving BSs in tiers $1,\ldots,k$.

Problem

5.1 Prove (5.3) as follows.

(1) From Theorem 2.4 and Problems 2.6 and 2.7, we have

$$\mathbb{P}\{\boldsymbol{\Gamma} > \boldsymbol{\gamma} \mid \boldsymbol{R} = \boldsymbol{r}\} = \begin{cases} 0, & \sum_{i=1}^{k} \gamma_i/(1+\gamma_i) \geq 1, \\ \det \mathbf{A}^{-1} \mathcal{L}_{W \mid \boldsymbol{R}=\boldsymbol{r}} \left(\boldsymbol{c}^\top \mathbf{A}^{-1} \boldsymbol{\rho} \right) \prod_{i=1}^{k} \frac{c_i}{\boldsymbol{c}^\top \mathbf{A}^{-1} \boldsymbol{e}_i^{(k)}}, & \sum_{i=1}^{k} \gamma_i/(1+\gamma_i) < 1, \end{cases} \tag{5.19}$$

where $\boldsymbol{c}$ is as defined in (5.4).

(2) From Problem 2.5, we have

$$\det \mathbf{A} = 1 - \mathbf{1}_k^\top \boldsymbol{\rho}, \quad \mathbf{A}^{-1} = \mathbf{I}_k + \frac{1}{1 - \mathbf{1}_k^\top \boldsymbol{\rho}} \boldsymbol{\rho} \mathbf{1}_k^\top.$$

Thus we have

$$\det \mathbf{A}^{-1} = \frac{1}{1 - \sum_{i=1}^{k} \gamma_i/(1+\gamma_i)}$$

and

$$\boldsymbol{c}^\top \mathbf{A}^{-1} \boldsymbol{e}_i^{(k)} = \boldsymbol{c}^\top \left[\boldsymbol{e}_i^{(k)} + \frac{1}{1 - \mathbf{1}_k^\top \boldsymbol{\rho}} \boldsymbol{\rho} \right] = c_i + \frac{\boldsymbol{c}^\top \boldsymbol{\rho}}{1 - \mathbf{1}_k^\top \boldsymbol{\rho}},$$

while

$$\boldsymbol{c}^\top \mathbf{A}^{-1} \boldsymbol{\rho} = \boldsymbol{c}^\top \left[1 + \frac{\mathbf{1}_k^\top \boldsymbol{\rho}}{1 - \mathbf{1}_k^\top \boldsymbol{\rho}} \right] \boldsymbol{\rho} = \frac{\boldsymbol{c}^\top \boldsymbol{\rho}}{1 - \mathbf{1}_k^\top \boldsymbol{\rho}}.$$

Substituting into (5.19) yields (5.3).

5.2.2 Application: camping probability in a macro-femto network

Consider an LTE macrocellular deployment with an overlaid cochannel femtocellular layer. In our notation, we have $n_{\text{tier}} = 3$ (macros, open femtos, and closed femtos, respectively) and $n_{\text{open}} = 2$ (macros and open femtos, respectively). The PPPs Φ_1, Φ_2, and Φ_3 for the macros, open femtos, and closed femtos, respectively, have intensities λ_1, λ_2, and λ_3. Equivalently, we may define Φ' to be the PPP of the femto BS locations (both open and closed), with intensity $\lambda' = \lambda_2 + \lambda_3$, wherein each femto BS of the PPP Φ' is selected to operate either in open mode with probability $p = \lambda_2/\lambda'$ or in closed mode with probability $1 - p = \lambda_3/\lambda'$, independent of the other femto BSs in Φ'. Note that the analysis applies to a specific instant in time (or, in LTE terminology, at each *transmit time interval* or TTI), but is also applicable to the dynamic scenario wherein the femto BSs in the PPP Φ' independently become configured in open mode with probability p and closed mode with probability $1 - p$ in each TTI.

For camping, the quantity of interest is the *reference symbol received power* (RSRP), i.e. the received power at the user location from the pilots or *reference symbols* (to borrow LTE terminology) that are broadcast by each BS in each tier. Thus

$$P = KP_{\mathrm{RS}}, \quad P' = K'P'_{\mathrm{RS}},$$

where P_{RS} and P'_{RS} are the RS transmit powers of the macro and femto (both open and closed) BSs, respectively. Note that we are using P and P' instead of P_1 and $P_2(= P_3)$, respectively, and similarly for K and K', the convention being that the unprimed quantities pertain to the macros and the primed ones pertain to the femtos.

In wireless system standards such as LTE, the camping criteria are expressed in terms of *reference symbol received quality* (RSRQ), defined as the ratio of RSRP to the total carrier received power (including both the desired signal and interference and thermal noise), measured over the same band as the RSRP measurement. In other words, $\mathrm{RSRQ} = \mathrm{SINR}/(1+\mathrm{SINR})$. We say that a user terminal (UE = user equipment in LTE terminology) can *camp on* the candidate serving macro [open femto] BS at distance r_1 [r_2] if $\mathrm{RSRQ}_1(r_1) > \theta_1$ [$\mathrm{RSRQ}(r_2) > \theta_2$] for some threshold θ_1 [θ_2], [1] which are equivalent to the conditions $\Gamma_1 > \gamma_1$ [$\Gamma_2 > \gamma_2$], where $\gamma_1 = (1-\theta_1)/\theta_1$ [$\gamma_2 = (1-\theta_2)/\theta_2$]. In LTE, $\gamma_1 = -4$ dB or higher (NTT DoCoMo, 2007). We define the UE to be in *outage* at this location if it cannot camp on any access point (macro Node B or femto BS operating in open mode, as the case may be), i.e. if there is no access point such that $\Gamma_1 > \gamma_1$ or $\Gamma_2 > \gamma_2$ at this UE location.

We assume that $\alpha_1 = \alpha_2 = \alpha_3 = \alpha$, say, and that the system is *interference limited*, i.e. $N_0 = 0$. We assume all fades are i.i.d. Rayleigh, so $G_{i,\alpha_i}(\cdot) = G_\alpha(\cdot)$, where $G_\alpha(\cdot)$ is given by (4.29). We also focus on user locations whose distance from the nearest macro BS is at least $r_{\min}$. Then, in (5.12), the lower limit of the integration over r_1, the distance of the user from the nearest macro BS, is not zero but $r_{\min}$. Equivalently, with the transformation of variables (5.14), the lower limit of integration over u_1 is not zero but $m_1 = \lambda_1 \pi r_{\min}^2$. We also observe that, for $\gamma_1 > 0, \gamma_2 > 0$,

$$\frac{\gamma_1}{1+\gamma_1} + \frac{\gamma_2}{1+\gamma_2} = 1 - \frac{1-\gamma_1\gamma_2}{(1+\gamma_1)(1+\gamma_2)} < 1 \Leftrightarrow \gamma_1\gamma_2 < 1.$$

From (5.7) and (5.13)–(5.17), the joint CCDF of the SIR at an arbitrary UE at a distance of at least $r_{\min}$ from its nearest macro BS is given by

$$\mathbb{P}\{\Gamma_1 > \gamma_1, \Gamma_2 > \gamma_2\} = \frac{(1+\gamma_1)(1+\gamma_2)}{1-\gamma_1\gamma_2} \int_{m_1}^{\infty} \int_0^{\infty} \exp\Bigg\{ -u_1 - u_2$$
$$- u_1 \left[\frac{(1+\gamma_1)(1+\gamma_2)}{1-\gamma_1\gamma_2} \left(\frac{\gamma_1}{1+\gamma_1} + \frac{\gamma_2}{1+\gamma_2} \left(\frac{u_2}{p\beta u_1} \right)^{\alpha/2} \right) \right]^{2/\alpha}$$
$$\times G_\alpha\left(\left[\frac{(1+\gamma_1)(1+\gamma_2)}{1-\gamma_1\gamma_2} \left(\frac{\gamma_1}{1+\gamma_1} + \frac{\gamma_2}{1+\gamma_2} \left(\frac{u_2}{p\beta u_1} \right)^{\alpha/2} \right) \right]^{-2/\alpha} \right)$$

[1] The LTE standard also requires $\mathrm{RSRP}_i(r_i) > P_{\min,i}$ for some $P_{\min,i}$, $i = 1, 2$, but these $P_{\min,i}$ are -121 dBm or lower (3GPP 2010, Sec. 5.2.3.2) and are nearly always satisfied for any macro BS or open femto BS at a UE location.

$$- u_2 \left[\frac{(1+\gamma_1)(1+\gamma_2)}{1-\gamma_1\gamma_2} \left(\frac{\gamma_2}{1+\gamma_2} + \frac{\gamma_1}{1+\gamma_1} \left(\frac{p\beta u_1}{u_2} \right)^{\alpha/2} \right) \right]^{2/\alpha}$$

$$\times G_\alpha \left(\left[\frac{(1+\gamma_1)(1+\gamma_2)}{1-\gamma_1\gamma_2} \left(\frac{\gamma_2}{1+\gamma_2} + \frac{\gamma_1}{1+\gamma_1} \left(\frac{p\beta u_1}{u_2} \right)^{\alpha/2} \right) \right]^{-2/\alpha} \right)$$

$$- \frac{1}{\operatorname{sinc}(2/\alpha)} \left[\frac{(1-p)^{\alpha/2}[\gamma_1(1+\gamma_2)(\beta u_1)^{\alpha/2} + \gamma_2(1+\gamma_1)(p^{-1}u_2)^{\alpha/2}]}{1-\gamma_1\gamma_2} \right]^{2/\alpha} \Bigg\} du_2\, du_1 \tag{5.20}$$

if $0 < \gamma_1\gamma_2 < 1$ and $\mathbb{P}\{\Gamma_1 > \gamma_1, \Gamma_2 > \gamma_2\} = 0$ if $\gamma_1\gamma_2 \geq 1$, where $p = \lambda_2/\lambda' = \lambda_2/(\lambda_2 + \lambda_3)$ is the fraction of femto BSs in *open access*, or OA, mode, and

$$\beta = \frac{\lambda'}{\lambda_1} \left(\frac{P'}{P} \right)^{2/\alpha} = \frac{\lambda'}{\lambda_1} \left(\frac{K' P'_{\text{RS}}}{K P_{\text{RS}}} \right)^{2/\alpha}. \tag{5.21}$$

The unconditional (area-wide) CCDF of the reference symbol SIR that the UE receives from its nearest macro BS, i.e. the probability that the UE can camp on its nearest macro BS, is given by (5.7) and (5.12) with $k = 1$ and the lower limit of integration over r_1 changed from 0 to $r_{\min}$:

$$\mathbb{P}\{\Gamma_1 > \gamma\} = \frac{\exp\left(-m_1 \left\{1 + \gamma^{2/\alpha} \left[G_\alpha(\gamma^{-2/\alpha}) + \beta/\operatorname{sinc}(2/\alpha)\right]\right\}\right)}{1 + \gamma^{2/\alpha}[G_\alpha(\gamma^{-2/\alpha}) + \beta/\operatorname{sinc}(2/\alpha)]}. \tag{5.22}$$

Note from (5.22) that the probability that a UE can camp on the nearest macro BS is entirely described by the quantities m_1 and β defined in (5.21). Observe that m_1 is the mean number of macro BSs within a disk of radius $r_{\min}$. Further, β is a function of the intensity ratios λ'/λ_1 and the power ratios $P'_{\text{RS}}/P_{\text{RS}}$. For a given m_1, all choices of femto BS intensity λ' and transmit power P'_{RS} relative to the corresponding macro BS values that yield the same β will give the same UE probability of camping on a macro BS. Thus, if the network operator chooses some values of m_1 and β (to satisfy a given macrocellular coverage requirement, say), then there are a variety of femtocellular overlays that will have the same macrocellular coverage. An extreme example is a femto deployment giving the femto BSs the same transmit power capabilities as the macro BSs and the same density. In terms of macrocellular coverage, if $\alpha = 4$, say, this femtocellular deployment is equivalent to another deployment where the femto BSs have 20 dB less power than the macro BSs and ten times the density of the macro BSs.

The probability that an arbitrary UE can camp on the nearest open femto BS can be written from (5.7) and (5.12) with $k = 1$ after swapping the labels of tiers 1 and 2. Some manipulation then yields

$$\mathbb{P}\{\Gamma_2 > \gamma\}$$
$$= \int_0^\infty \exp\left(-u\left\{1 + \gamma^{2/\alpha}\left[G_\alpha\left(\frac{1}{\gamma^{2/\alpha}}\right) + \frac{1/p - 1}{\operatorname{sinc}(2/\alpha)} + \frac{1}{p\beta} G_\alpha\left(\frac{m_1 p\beta}{\gamma^{2/\alpha} u}\right)\right]\right\}\right) du. \tag{5.23}$$

Again, it is seen that, for a given m_1 and fraction p of femto BSs that are in open mode, any changes to the powers or intensities of the femto BS or macro BS processes that leave β the same will lead to the same probability of coverage by an open femto BS.

Finally, the probability that a UE can camp on a BS (macro or open femto) is given by

$$\mathbb{P}(\{\Gamma_1 > \gamma\} \cup \{\Gamma_2 > \gamma\}) = \mathbb{P}\{\Gamma_1 > \gamma\} + \mathbb{P}\{\Gamma_2 > \gamma\} - \mathbb{P}\{\Gamma_1 > \gamma, \Gamma_2 > \gamma\}, \quad (5.24)$$

where the terms on the right-hand side are given by (5.22), (5.23), and (5.20), respectively. Note that the overall camping probability is also a function of m_1, β, p only, because this is true of each term in (5.24).

We set $\alpha = 4$ and $\gamma_1 = \gamma_2 = \gamma = 0.4$ (i.e. -4 dB). We set $K' = 0.00063$ (i.e. -32 dB) and $K = 12.6$ (i.e. 11 dB) consistent with an outdoor-to-indoor propagation model. In Figure 5.1, we plot the overall camping probability for an arbitrary UE, given by (5.24), vs. p, the fraction of open femto BSs, for several choices of r_{min} with $\lambda_1 = 0.1/\text{km}^2$, $\lambda' = 10/\text{km}^2$, $P_{\text{RS}} = 33$ dBm, and $P'_{\text{RS}} = 10$ dBm.

Note that, once we choose the desired minimum distance to a macro BS, r_{min}, to be some value in meters (m) or kilometers (km), a given value of m_1 on the plot corresponds to a macro BS density of $\lambda_1 = m_1/(\pi r_{\text{min}}^2)$, in units of BSs/m^2 or BSs/km^2. As previously discussed, a network operator considering a femtocellular overlay on an existing macrocellular deployment with intensity λ_1 can begin by choosing β to satisfy the macro BS camping requirement at distances $>r_{\text{min}}$ from the nearest macro BS (i.e. for a given m_1). Further, many different combinations of femto BS power and intensity

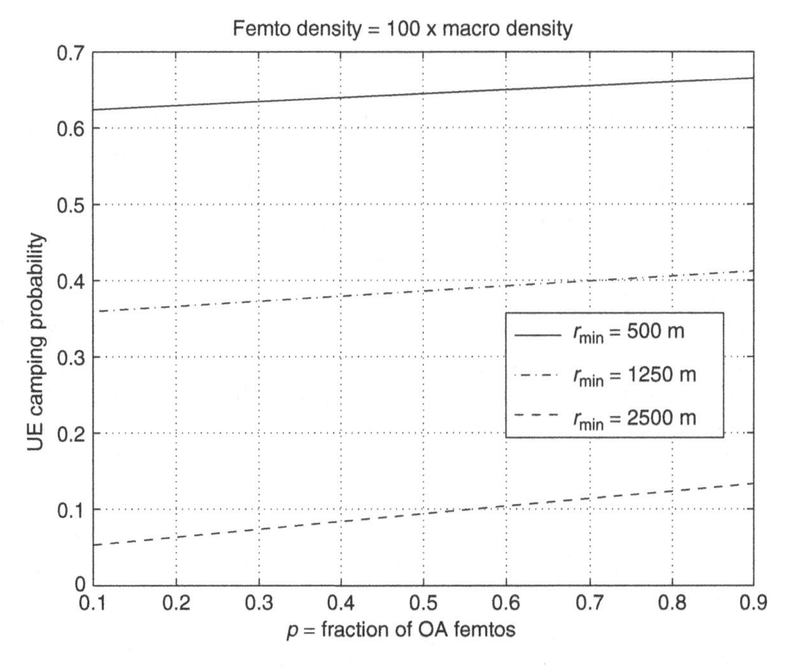

Figure 5.1 Plots of probability that the UE at distance $\geq r_{\text{min}}$ from the nearest macro BS can camp on either that macro BS or the nearest open femto BS, versus p, the fraction of OA femtos, for three values of r_{min}.

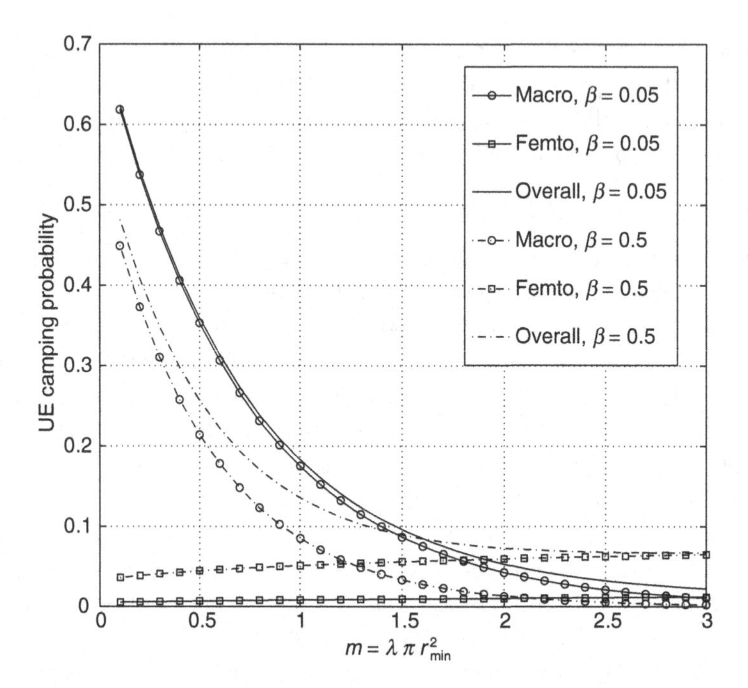

Figure 5.2 Plots of probability that the UE at distance $\geq r_{min}$ from the nearest macro BS can camp on that macro BS, the nearest open femto BS, or either, vs. m_1, the mean number of macro BSs within distance r_{min} of the UE location. We show the plots for $\beta = 0.05$ and 0.5. Here $p = 0.1$, meaning 10% of the femto BSs are in OA mode.

(relative to the macro BSs) could correspond to the same value of β. We see from (5.21) that, for the deployment with the given parameters, the value of $\beta = 0.05$. However, the same β could also correspond to the scenario $P'_{RS} = 23$ dBm and $\lambda' = 2.24/\text{km}^2$ (which is more characteristic of a macro-pico layout). The optimal choice from these combinations of the dimensionless parameters m_1 and β involves factors such as the overall camping probability and some factors beyond the scope of this book, such as the cost of femto and macro deployments. To illustrate the dependence of the overall camping probability (5.24) on p, we plot it versus m_1 for the same choices of β with $p = 0.1$ in Figure 5.2, and repeat with $p = 0.9$ in Figure 5.3. For completeness, we also include the macro BS camping probability (5.22) and the open femto BS camping probability (5.23) versus m_1 for the same choices of β and p in Figures 5.2 and 5.3.

From Figures 5.2 and 5.3, we observe the general trend that, for a given p and m_1, the open femto BS camping probability increases with β, whereas the macro camping probability decreases with β. This causes the overall camping probability to decrease and then increase, but the increase is so gradual that the curve is almost flat with respect to m_1. As, for a given macro BS deployment, m_1 is equivalent to the minimum distance of the UE from a macro BS, this insensitivity of the overall camping probability to the minimum distance is desirable. We note that the overall camping probability becomes flatter as p increases. For any camping probability curve, we observe that β needs to

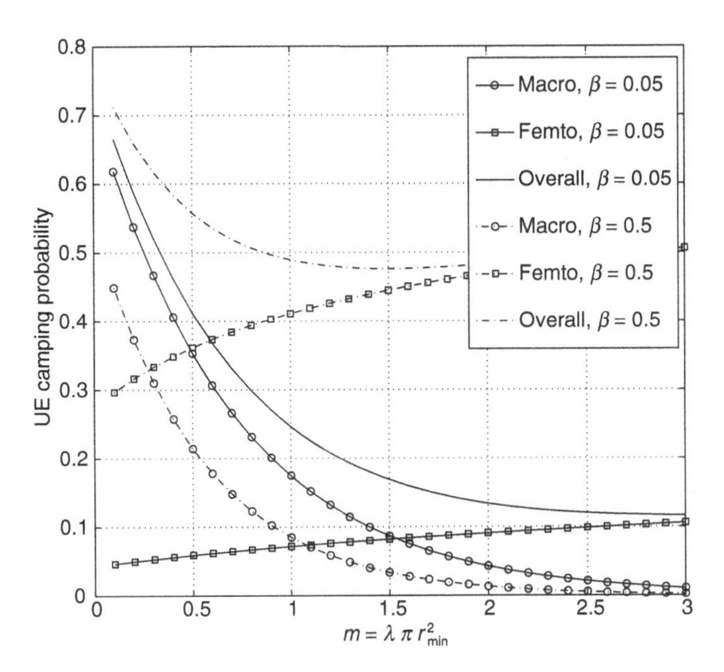

Figure 5.3 Same plots as in Figure 5.2, but with $p = 0.9$, corresponding to 90% of femto BSs operating in OA mode.

change by at least an order of magnitude to induce a significant change in the curve. Alternatively, plotting the camping probability vs. p in Figure 5.4, we note the increasing slope of the curves, i.e. greater sensitivity to p, as β increases.

5.2.3 Candidate serving BS in each tier is the one received most strongly at the user

Let us now consider the case where the candidate serving BS in each tier is the BS in that tier that is received with maximum instantaneous power at the user. In other words, this BS is also the maximum SINR (max-SINR) BS in that tier (see (4.48)). We assume, as previously, that the BS locations in the tiers are modeled by independent homogeneous PPPs. We also assume that fade attenuations on the links to a user location from all BSs in the network are independent, and that the fade attenuations on the links to the user from all BSs in a given tier are identically distributed. As previously, we focus on calculating the joint CCDF of the SINRs $\Gamma_1, \ldots, \Gamma_k$ at the user when it receives from the candidate serving BSs in the open tiers $1, \ldots, k$ for some $k \leq n_{\text{open}}$ (after relabeling the tiers if necessary):

$$\mathbb{P}\{\boldsymbol{\Gamma} > \boldsymbol{\gamma}\}, \quad \boldsymbol{\gamma} > \mathbf{0}_k, \quad \boldsymbol{\Gamma} = [\Gamma_1, \ldots, \Gamma_k]^\top, \; 1 \leq k \leq n_{\text{open}}. \tag{5.25}$$

From Theorem 4.1, we know that, for each tier (whether open or closed), the received powers at the user from the BSs in that tier with arbitrary i.i.d. fading on the links to the user form a one-dimensional PPP with the same intensity function as (i.e. equivalent to) the PPP of received powers from BSs located at points of another homogeneous PPP

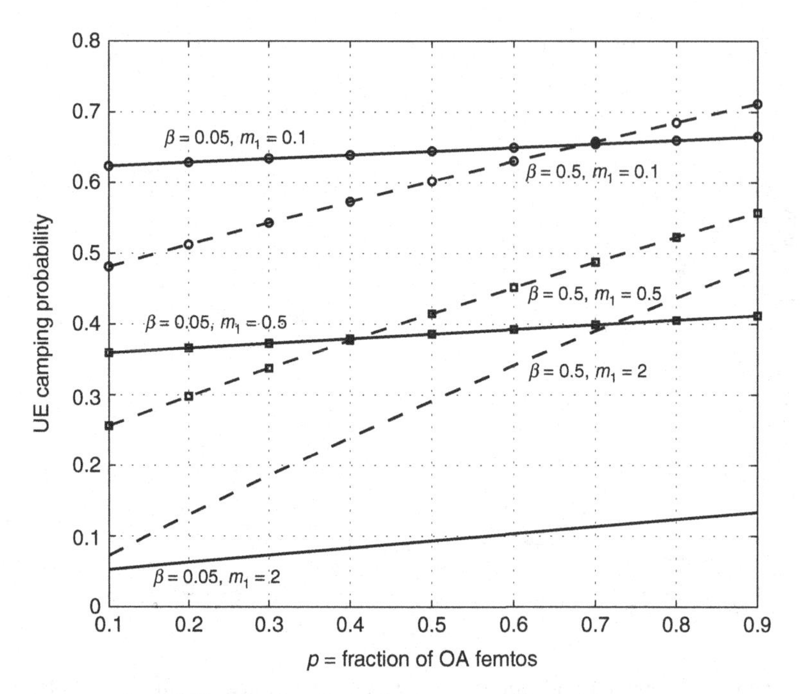

Figure 5.4 Probability that the UE (at distance $\geq r_{\text{min}}$ from the nearest macro BS, where r_{min} is such that m_1 is 0.1, 0.5, or 2) can camp on either that macro BS or the nearest open femto BS, vs. p, the fraction of OA fentocells, for $\beta = 0.05$ and 0.5.

with density modified according to (4.53), and with i.i.d. Rayleigh fading on the links to the user. Then the argument of Corollary 4.14 is easily generalized to state that not only are the marginal distributions of the SINRs from the BSs in the tiers the same as if the fades on the links from the BSs in those tiers were i.i.d. Rayleigh, but also the joint distribution of these SINRs is the same as if the fades on all links from all BSs in all tiers to the user were i.i.d. Rayleigh, and the densities of the PPPs representing the BSs in the tiers are modified according to (4.53) for each tier. In other words, so long as the BS locations in the tiers are points of independent homogeneous PPPs, the fade attenuations on the links to the user from all BSs in each tier are i.i.d. (with arbitrary continuous distribution), and these fade attenuations are independent (though not necessarily identically distributed) across the tiers, the joint CCDF (5.25) is the same as if all fade attenuations on all links (from all BSs in all tiers) to the user are i.i.d. Rayleigh, and the densities of the PPPs modeling the BS locations in the tiers are modified from the original densities according to the formula (4.53) for each tier. Without loss of generality, therefore, we shall assume that all fade attenuations on all links to the user are i.i.d. Rayleigh, and we calculate (5.25) for the case where the BS locations in the tiers are specified by independent homogeneous PPPs with density λ_i for tier i, $i = 1, \ldots, n_{\text{tier}}$.

The derivation of the joint CCDF (5.25) proceeds very similarly to that in Section 4.3.4. Denote the point pattern of the PPP for tier i by $\Phi_i = \{b_{i,1}, b_{i,2}, \ldots\}$, $i = 1, \ldots, n_{\text{tier}}$. Let $\Phi = \bigcup_{i=1}^{n_{\text{tier}}} \Phi_i$ be the point pattern of the superposition of the tier

PPPs. From Theorem 3.2, Φ is also a PPP. Similar to (4.54) and (4.56), we may write the joint CCDF as follows:

$$\begin{aligned}
&\mathbb{P}\{\Gamma_1 > \gamma_1, \ldots, \Gamma_k > \gamma_k\} \\
&= \sum_{n_1 \geq 1, \ldots, n_k \geq 1} \frac{(-1)^{n_1+\cdots+n_k-k}}{n_1! \cdots n_k!} \\
&\quad \times \sum_{\substack{b_{j,1},\ldots,b_{j,n_j} \in \Phi_j \\ j=1,\ldots,k}} \mathbb{P}\left(\bigcap_{i=1}^{k} \bigcap_{m=1}^{n_i} \left\{\frac{Y_{b_{i,m}}}{\sum_{b' \in \Phi \setminus \{b_{i,m}\}} Y_{b'} + N_0} > \gamma_i\right\}\right) \\
&= \sum_{n=k}^{\infty} \frac{(-1)^{n-k}}{n!} \sum_{\substack{n_1 \geq 1, \ldots, n_k \geq 1: \\ n_1+\cdots+n_k=n}} \binom{n}{n_1, \ldots, n_k} \sum_{b_{1,1} \in \Phi_1, \ldots, b_{1,n_1} \in \Phi_1} \cdots \sum_{b_{k,1} \in \Phi_k, \ldots, b_{k,n_k} \in \Phi_k} \\
&\quad \mathbb{P}\left(\bigcap_{i=1}^{k} \bigcap_{m=1}^{n_i} \left\{\frac{Y_{b_{i,m}}}{\sum_{b' \in \Phi \setminus \{b_{i,m}\}} Y_{b'} + N_0} > \gamma_i\right\}\right).
\end{aligned} \tag{5.26}$$

Basic notation

Before we proceed, let us establish some basic notation that will simplify the expressions we write in the future. For any vector $\boldsymbol{n}$ with k positive integer-valued entries $\boldsymbol{n} = [n_1, \ldots, n_k]^\top$, we write n in place of $\mathbf{1}_k^\top \boldsymbol{n} = n_1 + \cdots + n_k$ for simplicity. Also, we define $n_0 \equiv 0$ for convenience of notation in what follows. Given $\boldsymbol{n}$, define

$$\mathcal{N}_i = \{n_0 + \cdots + n_{i-1} + 1, \ldots, n_0 + \cdots + n_{i-1} + n_i\}, \quad i = 1, \ldots, k. \tag{5.27}$$

We now define the following function $h_{\boldsymbol{n}}$ of $\boldsymbol{x} = [x_1, \ldots, x_n]^\top \in \mathbb{R}^n$ and a countable subset $\mathcal{A} \subseteq \mathbb{R}_+$:

$$\begin{aligned}
h_{\boldsymbol{n}}(\boldsymbol{x}, \mathcal{A}) &\equiv \prod_{i=1}^{k} h_{n_i}\left(x_{n_0+\cdots+n_{i-1}+1}, \ldots, x_{n_0+\cdots+n_{i-1}+n_i}, \mathcal{A} \bigcup_{j \notin \mathcal{N}_i} \{x_j\}\right) \\
&= \prod_{i=1}^{k} \prod_{m \in \mathcal{N}_i} 1_{(\gamma_i, \infty)}\left(\frac{x_m}{\sum_{l \in \mathcal{N}_i \setminus \{m\}} x_l + \sum_{j \notin \mathcal{N}_i} x_j + \sum_{y \in \mathcal{A}} y + N_0}\right) \\
&= 1_{\mathcal{U}_{\boldsymbol{n}}\left(\sum_{y \in \mathcal{A}} y + N_0\right)}(\boldsymbol{x}),
\end{aligned} \tag{5.28}$$

where h_{n_i} is as defined in (4.58) with n and γ replaced by n_i and γ_i, respectively, and, for any $w > 0$, the set $\mathcal{U}_{\boldsymbol{n}}(w)$ is defined by (see Problem 5.2)

$$\begin{aligned}
\mathcal{U}_{\boldsymbol{n}}(w) &= \left\{\boldsymbol{x} \in \mathbb{R}^n : \mathbf{A}_{\boldsymbol{n}} \boldsymbol{x} > w \left[\rho_1 \mathbf{1}_{n_1}^\top, \ldots, \rho_k \mathbf{1}_{n_k}^\top\right]^\top\right\}, \\
\mathbf{A}_{\boldsymbol{n}} &= \mathbf{I}_n - \left[\rho_1 \mathbf{1}_{n_1}^\top, \ldots, \rho_k \mathbf{1}_{n_k}^\top\right]^\top \mathbf{1}_n^\top.
\end{aligned} \tag{5.29}$$

Finally, we also write

$$\boldsymbol{\rho_n} \equiv [\rho_{\boldsymbol{n},1}, \ldots, \rho_{\boldsymbol{n},n}]^\top = \left[\rho_1 \mathbf{1}_{n_1}^\top, \ldots, \rho_k \mathbf{1}_{n_k}^\top\right]^\top$$
$$= \Big[\underbrace{\rho_1, \ldots, \rho_1}_{n_1}, \underbrace{\rho_2, \ldots, \rho_2}_{n_2}, \ldots, \underbrace{\rho_k, \ldots, \rho_k}_{n_k}\Big]^\top \tag{5.30}$$

to distinguish it from our usual notation of $\boldsymbol{\rho} = [\rho_1, \ldots, \rho_k]^\top$.

Applying the extended Slivnyak–Mecke theorem

We may now resume the derivation of the joint CCDF from (5.26). As in Section 4.3.4, we denote the set of received powers $\tilde{\Phi}_i = \{Y_b : b \in \Phi_i\}$, $i = 1, \ldots, n_{\text{tier}}$, to be the one-dimensional inhomogeneous PPP with intensity function

$$\tilde{\lambda}_i(y) = 2\pi\lambda \int_0^\infty \frac{r^{1+\alpha_i}}{P_i} \exp\left(-\frac{y r^{\alpha_i}}{P_i}\right) \mathrm{d}r, \quad y \geq 0, \quad i = 1, \ldots, n_{\text{tier}}, \tag{5.31}$$

and let $\tilde{\Phi} = \cup_{i=1}^{n_{\text{tier}}} \tilde{\Phi}_i$ be the PPP obtained from the superposition of these PPPs. With this notation, we may rewrite (5.26) similarly to (4.61) as follows:

$$\mathbb{P}\{\Gamma_1 > \gamma_1, \ldots, \Gamma_k > \gamma_k\}$$
$$= \sum_{n=k}^{\infty} \frac{(-1)^{n-k}}{n!} \sum_{\substack{n_1 \geq 1, \ldots, n_k \geq 1: \\ n_1 + \cdots + n_k = n}} \binom{n}{n_1, \ldots, n_k} \mathbb{E} \sum_{\substack{y_1, \ldots, y_n \in \tilde{\Phi}: \\ \forall l \neq j,\, y_l \neq y_j \\ \forall m \in \mathcal{N}_i,\, y_m \in \tilde{\Phi}_i}} h_{\boldsymbol{n}}(\boldsymbol{y}, \tilde{\Phi} \setminus \{y_1, \ldots, y_n\}). \tag{5.32}$$

Next, we observe that (see Problem 5.3) Theorem 4.15 also holds for the superposed PPP $\tilde{\Phi}$ with the constraint that the first n_1 arguments of $h_{\boldsymbol{n}}$ in (5.32) belong to the PPP $\tilde{\Phi}_1$, the next n_2 arguments belong to $\tilde{\Phi}_2$, and so on, with the change that, in (4.62), the intensity function of the first n_1 variables of integration is $\tilde{\lambda}_1$ from (5.31), the intensity function of the next n_2 variables of integration is $\tilde{\lambda}_2$, and so on.

Applying (5.47), and with the usual definition of W as the thermal noise plus total received power at the user from all BSs in all tiers,

$$W = N_0 + \sum_{y \in \tilde{\Phi}} y,$$

we can follow the same sequence of steps as in the derivation of (4.66) to obtain

$$\mathbb{E} \sum_{\substack{y_1, \ldots, y_n \in \tilde{\Phi}: \\ \forall l \neq j,\, y_l \neq y_j \\ \forall m \in \mathcal{N}_i,\, y_m \in \tilde{\Phi}_i}} h_{\boldsymbol{n}}(\boldsymbol{y}, \tilde{\Phi} \setminus \{y_1, \ldots, y_n\})$$
$$= \int \cdots \int_{\mathbb{R}_{++}^n} \left[\prod_{i=1}^{k} \left(\frac{2\pi\lambda_i}{P_i}\right)^{n_i} \prod_{m \in \mathcal{N}_i} r_m^{1+\alpha_i}\right]$$
$$\mathbb{E}\left[\int_{\mathcal{U}_{\boldsymbol{n}}(W) \cap \mathbb{R}_{++}^n} \prod_{i=1}^{k} \prod_{m \in \mathcal{N}_i} \exp\left(-\frac{r_m^{\alpha_i}}{P_i} x_m\right) \mathrm{d}\boldsymbol{x}\right] \mathrm{d}r_1 \cdots \mathrm{d}r_n. \tag{5.33}$$

Given $\boldsymbol{n}$ and $r_1, \ldots, r_n$, let us define the independent exponential random variables

$$X_m(r_m) \sim \text{Exp}\left(\frac{P_i}{r_m^{\alpha_i}}\right), \quad m \in \mathcal{N}_i, \quad i = 1, \ldots, k,$$

and the vector $\boldsymbol{X}(r_1, \ldots, r_n) = [X_1(r_1), \ldots, X_n(r_n)]^\top$. We can then follow the same steps as in the derivation of (4.67) to obtain

$$\begin{aligned}\left(\prod_{i=1}^{k} \prod_{m \in \mathcal{N}_i} \frac{r_m^{\alpha_i}}{P_i}\right) \mathbb{E}\left[\int_{\mathcal{U}_n(W) \cap \mathbb{R}_{++}^n} \prod_{i=1}^{k} \prod_{m \in \mathcal{N}_i} \exp\left(-\frac{r_m^{\alpha_i}}{P_i} x_m\right) \mathrm{d}\boldsymbol{x}\right] \\ = \mathbb{E}\left[\mathbb{P}\{\boldsymbol{X}(r_1, \ldots, r_n) \in \mathcal{U}_{\boldsymbol{n}}(W) \cap \mathbb{R}_{++}^n \mid W\}\right] \\ = \mathbb{P}\left\{\mathbf{A}_{\boldsymbol{n}} \boldsymbol{X}(r_1, \ldots, r_n) > W \boldsymbol{\rho}_{\boldsymbol{n}}\right\}. \end{aligned} \tag{5.34}$$

From Theorem 2.4 and Problems 2.6 and 2.7, we know that

$$\mathbb{P}\left\{\mathbf{A}_{\boldsymbol{n}} \boldsymbol{X}(r_1, \ldots, r_n) > W \boldsymbol{\rho}_{\boldsymbol{n}}\right\} = 0 \text{ if } \mathbf{1}_n^\top \boldsymbol{\rho}_{\boldsymbol{n}} = \boldsymbol{n}^\top \boldsymbol{\rho} \geq 1$$

and, if $\boldsymbol{n}^\top \boldsymbol{\rho} < 1$,

$$\begin{aligned}&\mathbb{P}\left\{\mathbf{A}_{\boldsymbol{n}} \boldsymbol{X}(r_1, \ldots, r_n) > W \boldsymbol{\rho}_{\boldsymbol{n}}\right\} \\ &= \det \mathbf{A}_{\boldsymbol{n}}^{-1} \prod_{i=1}^{k} \prod_{m \in \mathcal{N}_i} \frac{r_m^{\alpha_i}/P_i}{\left[\frac{r_1^{\alpha_{\boldsymbol{n},1}}}{P_{\boldsymbol{n},1}}, \ldots, \frac{r_n^{\alpha_{\boldsymbol{n},n}}}{P_{\boldsymbol{n},n}}\right]^\top \mathbf{A}_{\boldsymbol{n}}^{-1} \boldsymbol{e}_m^{(n)}} \mathcal{L}_W\left(\left[\frac{r_1^{\alpha_{\boldsymbol{n},1}}}{P_{\boldsymbol{n},1}}, \ldots, \frac{r_n^{\alpha_{\boldsymbol{n},n}}}{P_{\boldsymbol{n},n}}\right]^\top \mathbf{A}_{\boldsymbol{n}}^{-1} \boldsymbol{\rho}_{\boldsymbol{n}}\right),\end{aligned} \tag{5.35}$$

where we have used the notation of (5.30) applied to the vectors $\boldsymbol{\alpha} = [\alpha_1, \ldots, \alpha_k]^\top$ and $\boldsymbol{P} = [P_1, \ldots, P_k]^\top$. Further, from (2.23) and (2.22),

$$\mathbf{A}_{\boldsymbol{n}}^{-1} = \mathbf{I}_n + \frac{1}{1 - \boldsymbol{n}^\top \boldsymbol{\rho}} \boldsymbol{\rho}_{\boldsymbol{n}} \mathbf{1}_n^\top, \quad \det \mathbf{A}_{\boldsymbol{n}} = 1 - \boldsymbol{n}^\top \boldsymbol{\rho}.$$

With the change of variables from $(r_1, \ldots, r_n)$ to $(u_1, \ldots, u_n)$, where

$$u_m = \left(\frac{\rho_i}{1 - \boldsymbol{n}^\top \boldsymbol{\rho}}\right)^{2/\alpha_i} \pi \lambda_i r_m^2, \; m \in \mathcal{N}_i, \quad i = 1, \ldots, k,$$

we can then write

$$\left[\frac{r_1^{\alpha_{\boldsymbol{n},1}}}{P_{\boldsymbol{n},1}}, \ldots, \frac{r_n^{\alpha_{\boldsymbol{n},n}}}{P_{\boldsymbol{n},n}}\right]^\top \mathbf{A}_{\boldsymbol{n}}^{-1} \boldsymbol{\rho}_{\boldsymbol{n}} = \frac{1}{1 - \boldsymbol{n}^\top \boldsymbol{\rho}} \sum_{i=1}^{k} \frac{\rho_i}{P_i} \sum_{m \in \mathcal{N}_i} r_m^{\alpha_i} = \sum_{i=1}^{k} \frac{\sum_{m \in \mathcal{N}_i} u_m^{\alpha_i/2}}{(\pi \lambda_i)^{\alpha_i/2} P_i},$$

while, for any $i = 1, \ldots, k$ and any $m \in \mathcal{N}_i$, we have

$$\begin{aligned}\left[\frac{r_1^{\alpha_{\boldsymbol{n},1}}}{P_{\boldsymbol{n},1}}, \ldots, \frac{r_n^{\alpha_{\boldsymbol{n},n}}}{P_{\boldsymbol{n},n}}\right]^\top \mathbf{A}_{\boldsymbol{n}}^{-1} \boldsymbol{e}_m^{(n)} &= \frac{r_m^{\alpha_i}}{P_i} + \frac{1}{1 - \boldsymbol{n}^\top \boldsymbol{\rho}} \sum_{j=1}^{k} \frac{\rho_j}{P_j} \sum_{l \in \mathcal{N}_j} r_l^{\alpha_j} \\ &= \frac{(1 - \boldsymbol{n}^\top \boldsymbol{\rho}) u_m^{\alpha_i/2}}{\rho_i (\pi \lambda_i)^{\alpha_i/2} P_i} + \sum_{j=1}^{k} \frac{\sum_{l \in \mathcal{N}_j} u_l^{\alpha_j/2}}{(\pi \lambda_j)^{\alpha_j/2} P_j}.\end{aligned}$$

Finally, note from (4.70) that the total received power W_j from the BSs in each tier $j = 1, \ldots, n_{\text{tier}}$ has the Laplace transform

$$\mathcal{L}_{W_j}(s) = \exp\left[-\pi\lambda_j \frac{(P_j s)^{2/\alpha_j}}{\text{sinc}(2/\alpha_j)}\right], \quad s > 0, \quad j = 1, \ldots, n_{\text{tier}},$$

so, from the independence of the fades across the tiers, we obtain the Laplace transform of W, the sum of the thermal noise and all received powers from all BSs in all tiers, to be

$$\mathcal{L}_W(s) = \exp\left[-sN_0 - \pi \sum_{j=1}^{n_{\text{tier}}} \frac{\lambda_j P_j^{2/\alpha_j}}{\text{sinc}(2/\alpha_j)} s^{2/\alpha_j}\right], \quad s > 0. \tag{5.36}$$

We therefore obtain

$$\left(\prod_{i=1}^{k} \prod_{m \in \mathcal{N}_i} \frac{r_m^{\alpha_i}}{P_i}\right) \mathbb{E}\left[\int_{\mathcal{U}_n(W) \cap \mathbb{R}_{++}^n} \prod_{i=1}^{k} \prod_{m \in \mathcal{N}_i} \exp\left(-\frac{r_m^{\alpha_i}}{P_i} x_m\right) \mathrm{d}\boldsymbol{x}\right] = \frac{1_{(-\infty,1)}(\boldsymbol{n}^\top \boldsymbol{\rho})}{1 - \boldsymbol{n}^\top \boldsymbol{\rho}}$$
$$\times \frac{\exp\left[-N_0 \sum_{i=1}^{k} \frac{\sum_{m \in \mathcal{N}_i} u_m^{\alpha_i/2}}{(\pi\lambda_i)^{\alpha_i/2} P_i} - \pi \sum_{j=1}^{n_{\text{tier}}} \frac{\lambda_j P_j^{2/\alpha_j}}{\text{sinc}(2/\alpha_j)} \left(\sum_{i=1}^{k} \frac{\sum_{m \in \mathcal{N}_i} u_m^{\alpha_i/2}}{(\pi\lambda_i)^{\alpha_i/2} P_i}\right)^{2/\alpha_j}\right]}{\prod_{i=1}^{k} \prod_{m \in \mathcal{N}_i} \left[1 + \frac{\rho_i (\pi\lambda_i)^{\alpha_i/2} P_i}{(1 - \boldsymbol{n}^\top \boldsymbol{\rho}) u_m^{\alpha_i/2}} \sum_{j=1}^{k} \frac{\sum_{l \in \mathcal{N}_j} u_l^{\alpha_j/2}}{(\pi\lambda_j)^{\alpha_j/2} P_j}\right]}. \tag{5.37}$$

Substituting (5.37) into (5.33) and then into (5.32), we finally obtain the following result (Mukherjee, 2011a).

THEOREM 5.4 *Suppose we are given n_{tier} tiers of BSs located at the points of independent homogeneous PPPs with densities $\nu_1, \ldots, \nu_{n_{\text{tier}}}$. Suppose the candidate serving BSs from the open tiers are chosen as the BSs received most strongly at the user. Then, with arbitrary distribution of the fade attenuations on the links to the user from the BSs in the tiers, independent on all links and identically distributed as some random variable H_i for all links from all BSs in tier i, $i = 1, \ldots, n_{\text{tier}}$, the joint CCDF of the SINRs at the user when receiving from the candidate serving BSs in open tiers $1, \ldots, k$ for any $k \leq n_{\text{open}}$ at any $\gamma_1 > 0, \ldots, \gamma_k > 0$ is given by*

$$\mathbb{P}\{\Gamma_1 > \gamma_1, \ldots, \Gamma_k > \gamma_k\}$$
$$= \sum_{n=k}^{\infty} \frac{(-1)^{n-k}}{n!} \sum_{\substack{n_1 \geq 1, \ldots, n_k \geq 1: \\ n_1 + \cdots + n_k = n \\ n_1\rho_1 + \cdots + n_k\rho_k < 1}} \binom{n}{n_1, \ldots, n_k} \frac{1}{1 - \boldsymbol{n}^\top \boldsymbol{\rho}} \prod_{i=1}^{k} \left(\frac{1 - \boldsymbol{n}^\top \boldsymbol{\rho}}{\rho_i}\right)^{2n_i/\alpha_i}$$

$$\times \int_{\mathbb{R}^n_{++}} \frac{\exp\left[-N_0 \sum_{i=1}^{k} \frac{\sum_{m\in\mathcal{N}_i} u_m^{\alpha_i/2}}{(\pi\lambda_i)^{\alpha_i/2} P_i} - \pi \sum_{j=1}^{n_{\text{tier}}} \frac{\lambda_j P_j^{2/\alpha_j}}{\text{sinc}(2/\alpha_j)} \left(\sum_{i=1}^{k} \frac{\sum_{m\in\mathcal{N}_i} u_m^{\alpha_i/2}}{(\pi\lambda_i)^{\alpha_i/2} P_i}\right)^{2/\alpha_j}\right]}{\prod_{i=1}^{k} \prod_{m\in\mathcal{N}_i} \left[1 + \frac{\rho_i (\pi\lambda_i)^{\alpha_i/2} P_i}{(1-\boldsymbol{n}^\top \boldsymbol{\rho}) u_m^{\alpha_i/2}} \sum_{j=1}^{k} \frac{\sum_{l\in\mathcal{N}_j} u_l^{\alpha_j/2}}{(\pi\lambda_j)^{\alpha_j/2} P_j}\right]} \mathrm{d}\boldsymbol{u}, \tag{5.38}$$

where $\lambda_i = \nu_i \, \mathbb{E}[H_i^{2/\alpha_i}]/\Gamma(1+2/\alpha_i)$, $i = 1, \ldots, n_{\text{tier}}$, *and* $\rho_i = \gamma_i/(1+\gamma_i)$, $i = 1, \ldots, k$.

Remark 5.3 It is also clear from (5.38) that the joint distribution of SINRs when the candidate serving BSs in the open tiers are chosen as the ones received most strongly at the user is a function of only the thermal noise power N_0 and the deployment parameters in the n_{tier} combinations $\lambda_i^{\alpha_i/2} K_i P_i^{\text{tx}}$, $i = 1, \ldots, n_{\text{tier}}$, exactly as when the candidate serving BSs are chosen as the ones nearest to the user (see Theorem 5.3).

Remark 5.4 Note that the right side of (5.38) may appear to be an infinite sum, but it is in fact finite, because for any n the multiple-sum over the partitions $n_1, \ldots, n_k$ of n is a finite sum over those positive integers $n_1, \ldots, n_k$ that satisfy both $n_1 + \cdots + n_k = n$ and $n_1\rho_1 + \cdots + n_k\rho_k < 1$. In other words, the multiple-sum is at most over $n_1, \ldots, n_k$ such that $1 \le n_i < \min\{n-k+1, \lceil \rho_i^{-1} \rceil - 1\} \le \lceil \rho_i^{-1} \rceil - 1$, $i = 1, \ldots, k$.

Remark 5.5 Finally, as with the derivation of (4.73) in the single-tier deployment, if we are interested in the SINR distribution at a user located at a minimum distance of $d_{\text{min},i}$ from all BSs of tier i, $i = 1, \ldots, k$, the lower limit of integration in (5.38) changes from zero to $u_{\text{min},i} \equiv \pi\lambda_i d_{\text{min},i}^2$ for all u_m, $m \in \mathcal{N}_i$, $i = 1, \ldots, k$, and $\mathcal{L}_W(s)$ changes from (5.36) to

$$\exp\left\{-sN_0 - \pi \sum_{i=1}^{k} \lambda_i (P_i s)^{2/\alpha_i} G_{\alpha_i}\left(\frac{d_{\text{min},i}^2}{(P_i s)^{2/\alpha_i}}\right) - \pi \sum_{j=k+1}^{n_{\text{tier}}} \frac{\lambda_j (P_j s)^{2/\alpha_j}}{\text{sinc}(2/\alpha_j)}\right\}, \quad s > 0,$$

where $G_\alpha(\cdot)$ is given by (4.29). Like (4.73), this expression is only applicable to i.i.d. Rayleigh fading on all links (see Remark 4.22).

Special case: N_0=0, same α for all tiers

When thermal noise can be neglected and the path-loss exponent is equal on the links to the user from all BSs in all tiers, we have the following simplification of (5.38).

THEOREM 5.5 *Under the hypotheses of Theorem 5.4, if the path-loss exponents α_i are equal (to α, say) for all $i = 1, \ldots, n_{\text{tier}}$, then the joint CCDF of the SIR at the user when receiving from the candidate serving BSs in open tiers $1, \ldots, k$ for any $k \le n_{\text{open}}$ depends only on the k combinations of the system deployment parameters defined by*

$$\beta_i = \frac{\lambda_i P_i^{2/\alpha_i}}{\sum_{j=1}^{n_{\text{tier}}} \lambda_j P_j^{2/\alpha_j}}, \quad i = 1, \ldots, k, \tag{5.39}$$

and, for any $\gamma_1 > 0, \ldots, \gamma_k > 0$*, is given by*

$$N_0 = 0,\ \alpha_1 = \cdots = \alpha_{n_{\mathrm{tier}}} = \alpha \Rightarrow \mathbb{P}\{\Gamma_1 > \gamma_1, \ldots, \Gamma_k > \gamma_k\}$$
$$= \sum_{n=k}^{\infty} \frac{(-1)^{n-k}[\mathrm{sinc}(2/\alpha)]^n}{n}$$
$$\times \sum_{\substack{n_1 \geq 1, \ldots, n_k \geq 1 \\ n_1 + \cdots + n_k = n \\ n_1\rho_1 + \cdots + n_k\rho_k < 1}} \binom{n}{n_1, \ldots, n_k} (1 - \boldsymbol{n}^\top \boldsymbol{\rho})^{2n/\alpha - 1} T_{\boldsymbol{n}}(\boldsymbol{\rho}, \alpha) \prod_{i=1}^{k} \left(\frac{\beta_i}{\rho_i^{2/\alpha}}\right)^{n_i}, \tag{5.40}$$

where if $n > 1$ $T_{\boldsymbol{n}}(\boldsymbol{\rho}, \alpha)$ *is given by*

$$T_{\boldsymbol{n}}(\boldsymbol{\rho}, \alpha) = \left(\frac{2}{\alpha}\right)^{n-1} \int_0^1 \mathrm{d}v_1 \int_0^{1-v_1} \mathrm{d}v_2 \cdots \int_0^{1-\sum_{m=1}^{n-2} v_m} \mathrm{d}v_{n-1}$$
$$\left[\prod_{m=1}^{n-1} \frac{v_m^{2/\alpha}}{v_m + \rho_{\boldsymbol{n},m}/(1 - \boldsymbol{n}^\top \boldsymbol{\rho})}\right] \frac{(1 - v_1 - \cdots - v_{n-1})^{2/\alpha}}{(1 - v_1 - \cdots - v_{n-1}) + \rho_{\boldsymbol{n},n}/(1 - \boldsymbol{n}^\top \boldsymbol{\rho})}, \tag{5.41}$$

and if $n = 1$ *(i.e. if* $k = 1$ *and* $n_1 = 1$*)*

$$T_1(\rho_1, \alpha) = \frac{1}{1 + \rho_1/(1 - \rho_1)} = 1 - \rho_1 = \frac{1}{1 + \gamma_1}.$$

Proof A straightforward extension of the method outlined in Problem 4.15. □

Observe that if $\alpha_1 = \cdots = \alpha_{n_{\mathrm{tier}}} = \alpha$, then, from (4.6), Corollary 4.14, and Theorem 3.2, the superposition PPP $\tilde{\Psi} = \cup_{i=1}^{n_{\mathrm{tier}}} \tilde{\Psi}$ of received powers from all BSs is equivalent to the one-dimensional PPP of received powers from BSs located at the points of a homogeneous *single-tier* PPP deployment Φ^*, all transmitting with unit power and with i.i.d. Rayleigh fading on all links to the user, where the density of Φ^* is given by

$$\lambda^* = \sum_{i=1}^{n_{\mathrm{tier}}} \lambda_i P_i^{2/\alpha}\, \mathbb{E}[H_i^{2/\alpha}].$$

Further, Theorem 3.3 then says that we can distinguish the original n_{tier} tiers from among this new equivalent single tier Φ^* of BSs by assigning a "color" to each tier, such that a BS in Φ^* has "color" i (i.e. belongs to tier i) with retention probability (5.39).

It follows that the marginal CCDF of the SINR from a single tier $i \in \{1, \ldots, n_{\mathrm{open}}\}$ of the original superposition PPP Φ, given by $\mathbb{P}\{\Gamma_i > \gamma\}$, can be calculated as the sum over all $n \geq 1$ of the probability that (a) the SINR at the user when receiving from n BSs of the new single-tier deployment Φ^* exceeds γ, and that (b) these n BSs all have "color" i. Now, from (4.56), (4.61), and the subsequent derivations leading up to (4.74), we see that the probability that the SINR at the user when receiving from exactly n BSs of the single-tier deployment is only non-zero for $n \leq \lceil \gamma^{-1} \rceil$, and is given by

$$\frac{[\mathrm{sinc}(2/\alpha)]^n}{n(1 - n\rho)} \left(\frac{1}{\rho} - n\right)^{2n/\alpha} T_n(\rho, \alpha), \tag{5.42}$$

where $T_n(\rho, \alpha)$ is given by (4.75) and $\rho = \gamma/(1+\gamma)$. The expression (5.42) is the probability of event (a) just given, while the probability of event (b) is simply β_i^n. Thus we obtain the *marginal* CCDF of the SINR at the user when receiving from the tier i as

$$N_0 = 0,\ \alpha_1 = \cdots = \alpha_{n_{\text{tier}}} = \alpha \Rightarrow \mathbb{P}\left\{\Gamma_i > \gamma \equiv \frac{\rho}{1-\rho}\right\}$$
$$= \sum_{n=1}^{\lceil \rho^{-1} \rceil - 1} \frac{(-1)^{n-1}[\operatorname{sinc}(2/\alpha)]^n}{n(1-n\rho)} \left(\frac{1}{\rho} - n\right)^{2n/\alpha} T_n(\rho,\alpha)\, \beta_i^n, \quad i = 1, \ldots, n_{\text{open}}. \tag{5.43}$$

Of course, (5.43) can be derived directly from (5.40) by substituting $k = 1$, $n_1 = n$, and Γ_i, γ in place of Γ_1, γ_1, respectively. Note that we cannot obtain $\mathbb{P}\{\Gamma_i > \gamma_i\}$ just by setting $\gamma_j = 0$ for $j \neq i$ in (5.40) because our derivation of the joint CCDF requires $\gamma_1, \ldots, \gamma_k$ to be strictly positive in (5.40). Further, if $\gamma \geq 1$ in (5.43), then only the $n = 1$ term survives, and we obtain

$$N_0 = 0,\ \alpha_1 = \cdots = \alpha_{n_{\text{tier}}} = \alpha \Rightarrow \mathbb{P}\{\Gamma_i > \gamma\}$$
$$= \beta_i \frac{\operatorname{sinc}(2/\alpha)}{\gamma^{2/\alpha}}, \quad \gamma \geq 1, \quad i = 1, \ldots, n_{\text{open}}. \tag{5.44}$$

Observe that $T_{\boldsymbol{n}}(\boldsymbol{\rho}, \alpha)$ in (5.41) reduces to $T_n(\rho, \alpha)$ in (4.75) when $k = 1 = n_{\text{open}} = n_{\text{tier}}$ and $\boldsymbol{n} = n_1 = n$, $\boldsymbol{\rho} = \rho_1 = \rho$. The following result presents another interesting case where $T_{\boldsymbol{n}}(\boldsymbol{\rho}, \alpha)$ becomes $T_n(\rho, \alpha)$.

COROLLARY 5.6 *In Theorem 5.5, suppose* $\gamma_1 = \cdots = \gamma_k \equiv \gamma$, *say, and define* $\rho = \gamma/(1+\gamma)$. *If* $\lceil \rho^{-1} \rceil > k$, *then*

$$N_0 = 0,\ \alpha_1 = \cdots = \alpha_{n_{\text{tier}}} = \alpha \Rightarrow \mathbb{P}\{\Gamma_1 > \gamma, \ldots, \Gamma_k > \gamma\}$$
$$= \sum_{n=k}^{\lceil \rho^{-1} \rceil - 1} \frac{(-1)^{n-k}[\operatorname{sinc}(2/\alpha)]^n}{n(1-n\rho)} \left(\frac{1}{\rho} - n\right)^{2n/\alpha} T_n(\rho,\alpha)$$
$$\times \sum_{\substack{n_1 \geq 1, \ldots, n_k \geq 1 \\ n_1 + \cdots + n_k = n}} \binom{n}{n_1, \ldots, n_k} \prod_{i=1}^{k} \beta_i^{n_i}, \tag{5.45}$$

where $T_n(\rho, \alpha)$ *is given by* (4.75), *otherwise* $\mathbb{P}\{\Gamma_1 > \gamma, \ldots, \Gamma_k > \gamma\} = 0$.

Proof Verify that if $\rho_1 = \cdots = \rho_k = \rho$, i.e. if $\boldsymbol{\rho} = \rho \mathbf{1}_k$, then, for any vector $\boldsymbol{n} = [n_1, \ldots, n_k]^\top$ with positive integer-valued entries, $T_{\boldsymbol{n}}(\boldsymbol{\rho}, \alpha)$ in (5.41) reduces to $T_n(\rho, \alpha)$ in (4.75). The result follows immediately from (5.40). □

Alternatively, $\mathbb{P}\{\Gamma_1 > \gamma, \ldots, \Gamma_k > \gamma\}$ can be calculated as the sum over all $n \geq k$ of the probability that (a) the SINR at the user when receiving from each of n BSs in Φ^* exceeds γ, and that (b) at least one of these BSs has "color" i for each $i = 1, \ldots, k$. From (4.74), the probability that the SINR at the user exceeds γ from exactly n BSs in Φ^* ($n \geq k$) is (5.42). This is the probability of (a). Further, given that there are $n \geq k$ BSs in Φ^* such that the SINR at the user when receiving from each of them exceeds γ, the probability that there is at least one BS with each of the colors $1, \ldots, k$ from among these n BSs is just the sum of the appropriate probabilities over all ways to partition n into k positive integers, which may be calculated as follows. For each such partition $n_1, \ldots, n_k$ of n, the appropriate probability that there are n_1 BSs with color $1, \ldots, n_k$ BSs with color k is $\binom{n}{n_1,\ldots,n_k}\beta_1^{n_1}\cdots\beta_k^{n_k}$. In other words, the probability that there is at least one BS with each of the colors $1, \ldots, k$ from among the n BSs in Φ^* that are received at the user with SINR exceeding γ is

$$\sum_{\substack{n_1\geq 1,\ldots,n_k\geq 1\\ n_1+\cdots+n_k=n}} \binom{n}{n_1,\ldots,n_k} \beta_1^{n_1}\cdots\beta_k^{n_k}. \tag{5.46}$$

This is the probability of event (b). Combining (5.42) and (5.46) yields (5.45).

Problems

5.2 Prove (5.29). *Hint*: Follow the same steps as in the derivation of (4.59).

5.3 Suppose $\Psi_1, \ldots, \Psi_k$ are independent PPPs with intensity functions $\nu_1, \ldots, \nu_k$, respectively, and $\Psi = \Psi' \cup \left(\cup_{i=1}^k \Psi_i\right)$ is the superposition of these PPPs and another independent PPP Ψ'. Given $\boldsymbol{n}$, prove, using Theorem 4.15, that, for any function g of $\boldsymbol{x} \in \mathbb{R}^n$ and a countable subset of $\mathbb{R}$, we have

$$\begin{aligned}\mathbb{E} \sum_{\substack{y_1,\ldots,y_n\in\Psi:\\ \forall l\neq j,\ y_l\neq y_j\\ \forall m\in\mathcal{N}_i,\ y_m\in\Psi_i}} g(\boldsymbol{y}, \Psi \setminus \{y_1,\ldots,y_n\})\\ = \int_{\mathbb{R}^n} \mathbb{E}[g(x_1,\ldots,x_n,\Psi)] \prod_{i=1}^{k}\prod_{m\in\mathcal{N}_i} \nu_i(x_m)\mathrm{d}x_m\,.\end{aligned} \tag{5.47}$$

5.2.4 Application: coverage probability in an HCN

Introduction

In Section 5.2.2 we studied the probability that an arbitrarily located user (UE in LTE terminology) in a multi-tier macro-femto HCN can *camp* on at least one accessible tier (macro or open access femto), where by "camp" we mean that the UE has sufficient SINR on the *pilot* channels to be able to identify the candidate serving BSs in each tier. In the present section, we apply a similar analysis to the probability of *coverage* of an arbitrarily located UE in a multi-tier HCN, where "coverage" essentially means that the UE has sufficient SINR on the *traffic* or data channels from the candidate serving BSs in the accessible tiers to be able to support some desired data rate *on those tiers*. Thus, the probability of coverage of the UE is the probability that the SINR when receiving from the candidate serving BSs of one of more open tiers exceeds some corresponding threshold. It is important to realize that the network may be designed with the goal of preferentially supporting certain services (with corresponding data rates) on certain tiers. For example, a macro-pico HCN may be designed with the goal of offering only a low base data rate when users are served by macro BSs, but a significantly higher data rate when they are served by femto BSs, which have smaller ranges because of their lower transmit powers. In this case, the SINR threshold for coverage by the pico tier should be higher than that for coverage by the macro tier.

In this section, we assume for coverage probability calculations that the candidate serving BSs from the open tiers are all chosen as the most strongly received BSs from those tiers at the UE, so, if the UE is served by a certain tier, we are certain that it has the highest SINR possible when served by any BS in that tier. We call this candidate serving BS selection criterion the max-SINR rule.

Note that we have not said anything yet about the selection criterion for the overall serving BS, i.e. the tier (and corresponding candidate serving BS) that is chosen to actually transmit to the UE. One obvious criterion would be simply to apply the max-SINR rule again to choose the candidate serving BS that is received most strongly at the UE. However, this is not the only criterion possible, and it is in fact not favored in HCN deployments, where usually some form of *biasing* is used to *offload* traffic from the macro to the other tier(s) in order to satisfy some additional design criteria, such as the desired fraction of overall UEs served by each tier. The dependence of coverage probability on an arbitrary overall serving BS selection criterion will be explored in this section, while the distribution of the marginal SINR from the selected overall serving BS under various selection criteria will be studied in Section 5.3. First, however, we need to establish some basic definitions and notation.

Definitions and notation

A given UE is said to be *covered by tier* $i \in \{1, \ldots, n_{\text{open}}\}$ if $\Gamma_i > \gamma_i$, where Γ_i is the SINR at this UE when it receives from the strongest BS in tier i, and $\gamma_i > 0$ is some *fixed threshold* value for tier i, $i = 1, \ldots, n_{\text{open}}$. A given UE is said to be *covered* if it is covered by some tier $i \in \{1, \ldots, n_{\text{open}}\}$. The *probability of coverage* P_{cov} is the probability that an arbitrary UE is covered:

$$P_{\text{cov}} = \mathbb{P}\left(\bigcup_{i=1}^{n_{\text{open}}} \{\Gamma_i > \gamma_i\}\right). \tag{5.48}$$

A given UE is said to be *covered under* a particular overall serving BS selection criterion (e.g. max-SINR) applied to the n_{open} candidate serving BSs (themselves chosen from the n_{open} open tiers by the max-SINR rule applied to each of these tiers) if the UE is covered by the tier I selected under this overall serving BS selection criterion. The *probability of coverage under the overall serving BS selection criterion* $P_{\text{cov}}(I)$ is defined as the probability that an arbitrary UE is covered under this selection criterion:

$$P_{\text{cov}}(I) = \mathbb{P}\{\Gamma_I > \gamma_I\}.$$

For later use, we also define the max-SINR rule for overall serving BS selection (from among the n_{open} candidate serving BSs) as that which yields the serving tier I^* given by

$$I^* = \arg \max_{1 \leq i \leq n_{\text{open}}} \Gamma_i. \tag{5.49}$$

In this section, we make frequent use of identities between indicator functions of sets. Since the arguments of the indicator functions are not used in this section, we drop them for convenience. This lets us write $1_{\mathcal{A}}$ instead of $1_{\mathcal{A}}(\cdot)$, for example. For greater legibility, we abuse the notation slightly and write $1(\mathcal{A})$ instead of $1_{\mathcal{A}}$ in the remainder of this section to denote the indicator function of the set $\mathcal{A}$.

Results on probability of coverage

We begin with the following result.

LEMMA 5.7 *With the above definitions, the probability of coverage P_{cov} is an upper bound on the probability of coverage $P_{\text{cov}}(I)$ under any serving BS selection criterion:*

$$P_{\text{cov}}(I) = \mathbb{P}\{\Gamma_I > \gamma_I\} \leq \mathbb{P}\left(\bigcup_{i=1}^{n_{\text{open}}} \{\Gamma_i > \gamma_i\}\right) = P_{\text{cov}}. \tag{5.50}$$

Proof This follows immediately from the fact that $\{\Gamma_I > \gamma_I\} \subseteq \cup_{i=1}^{n_{\text{open}}} \{\Gamma_i > \gamma_i\}$. □

We are interested in cases where the probability of coverage under the max-SINR rule for overall serving BS selection equals the probability of coverage P_{cov}, i.e., the event $\left(\cup_{i=1}^{n_{\text{open}}} \{\Gamma_i > \gamma_i\}\right)$ that the UE is covered is equivalent to the event $\{\Gamma_{I^*} > \gamma_{I^*}\}$ that it is covered under the max-SINR rule for overall serving BS selection. Note that for arbitrary thresholds $\gamma_1, \ldots, \gamma_{n_{\text{open}}}$, this is not true in general, as the following example demonstrates: consider a co-channel two-tier macro-pico HCN where both tiers are open, with 1 labeling the macro tier and 2 the pico tier, and corresponding SINR thresholds for coverage γ_1 and γ_2, with $\gamma_2 > \gamma_1$. Suppose that at the UE, the SINR from the candidate serving macro BS satisfies $\Gamma_1 > \gamma_1$, so the UE is covered because it is covered by the macro tier. However, suppose that the SINR from the candidate serving pico BS is higher than that from the candidate serving macro BS, i.e., $\Gamma_2 > \Gamma_1$, but owing to interference from the macro tier, this SINR from the candidate serving pico BS is below the threshold for pico coverage: $\Gamma_2 < \gamma_2$. In other words, we have described a scenario where $\gamma_2 > \Gamma_2 > \Gamma_1 > \gamma_1$, which can happen with finite probability. Then the max-SINR rule for overall serving BS selection would select the UE to be served by the pico tier, but then this UE would not be covered by the selected serving tier because $\Gamma_2 < \gamma_2$.

The next theorem provides some cases where the upper bound in (5.50) is achieved by the max-SINR rule for serving BS selection.

THEOREM 5.8 *With the hypotheses of Lemma 5.7, the max-SINR rule for serving BS selection achieves the upper bound in* (5.50), *i.e.,*

$$P_{\text{cov}}(I^*) = P_{\text{cov}}, \tag{5.51}$$

in each of the following cases:

(1) $n_{\text{open}} = 1$, *i.e., there is only one open tier;*
(2) $n_{\text{open}} > 1$ *and all coverage thresholds are equal:* $\gamma_1 = \cdots = \gamma_{n_{\text{open}}}$;
(3) $n_{\text{open}} > 1$ *and the coverage thresholds* $\gamma_1, \ldots, \gamma_{n_{\text{open}}}$ *are not necessarily equal, but*

$$\gamma_i > 1, \quad i = 1, \ldots, n_{\text{open}}. \tag{5.52}$$

Proof The proof when $n_{\text{open}} = 1$ is trivial, because then $I^* \equiv 1$. The proof when the coverage thresholds are equal (to γ, say) is also trivial, because then $\gamma_{I^*} \equiv \gamma$ whereas $\Gamma_{I^*} \equiv \max_{1 \le i \le n_{\text{open}}} \Gamma_i$, and $\{\max_{1 \le i \le n_{\text{open}}} \Gamma_i > \gamma\} \equiv \cup_{i=1}^{n_{\text{open}}} \{\Gamma_i > \gamma\}$ by definition. The proof of (5.51) when (5.52) holds is as follows. Consider the set of SINRs at the UE when receiving from the BSs in tiers $1, \ldots, n_{\text{open}}$. We know that at most one of these SINRs exceeds 1 (see Problem 4.18). It follows that if (5.52) holds, then the UE is covered by at most one BS in the HCN. In terms of indicator functions, and employing the notation described above, we therefore have with probability 1,

$$1\left(\bigcup_{i=1}^{n_{\text{open}}} \{\Gamma_i > \gamma_i\}\right) = \sum_{i=1}^{n_{\text{open}}} 1\{\Gamma_i > \gamma_i\}. \tag{5.53}$$

Further, if (5.52) holds, the serving BS must be the max-SINR BS (because only the SINR from this BS in each tier exceeds the required threshold for coverage in that tier), which means that if the serving tier is $i \in \{1, \ldots, n_{\text{open}}\}$, then we have $I^* = i$, where I^* is the max-SINR tier defined as in (5.49). It then follows that with probability 1,

$$1\{\Gamma_i > \gamma_i\} = 1\{\Gamma_i > \gamma_i\}1\{I^* = i\} = 1\{\Gamma_i > \gamma_i, I^* = i\}, \quad i = 1, \ldots, n_{\text{open}}. \tag{5.54}$$

Combining this with (5.53), we obtain

$$P_{\text{cov}} = \sum_{i=1}^{n_{\text{open}}} \mathbb{P}\{\Gamma_i > \gamma_i\} = \sum_{i=1}^{n_{\text{open}}} \mathbb{P}\{\Gamma_i > \gamma_i, I^* = i\} = \mathbb{P}\{\Gamma_{I^*} > \gamma_{I^*}\} = P_{\text{cov}}(I^*). \tag{5.55}$$

This proves (5.51). □

It turns out that Theorem 5.8 also has a converse, showing that for these three cases, the upper bound in (5.50) is only achieved by a serving BS selection rule that coincides with the max-SINR rule when coverage under the latter rule occurs.

THEOREM 5.9 *Under the hypotheses of Theorem 5.8, and for each of the cases therein, we have*

$$P_{\text{cov}}(I) \le P_{\text{cov}}(I^*) = P_{\text{cov}} \tag{5.56}$$

for any serving BS selection criterion giving serving tier I, with strict inequality if this serving BS selection criterion does not always select the max-SINR BS when coverage occurs, i.e., if $\mathbb{P}\{I = I^* \mid \Gamma_{I^*} > \gamma_{I^*}\} < 1$.

Proof The proof for the first two cases in Theorem 5.8 is trivial because there "coverage" is identical to "coverage under the max-SINR rule" by definition. The proof of (5.56) for the third case is as follows. From the discussion leading up to (5.54), we note that under the assumption (5.52), we have with probability 1

$$1\{\Gamma_i > \gamma_i, I = i, I \neq I^*\} = 0, \quad i = 1, \ldots, n_{\text{open}}.$$

This in turn implies that with probability 1,

$$\begin{aligned}
1\{\Gamma_I > \gamma_I\} &= \sum_{i=1}^{n_{\text{open}}} 1\{\Gamma_i > \gamma_i, I = i\} \\
&= \sum_{i=1}^{n_{\text{open}}} \left[1\{\Gamma_i > \gamma_i, I = i, I = I^*\} + 1\{\Gamma_i > \gamma_i, I = i, I \neq I^*\}\right] \\
&= \sum_{i=1}^{n_{\text{open}}} 1\{\Gamma_i > \gamma_i, I^* = i, I^* = I\} \\
&= 1\{\Gamma_{I^*} > \gamma_{I^*}, I = I^*\},
\end{aligned}$$

which gives

$$P_{\text{cov}}(I) = \mathbb{P}\{\Gamma_I > \gamma_I\} = \mathbb{P}\{\Gamma_{I^*} > \gamma_{I^*}, I = I^*\} = P_{\text{cov}}(I^*)\,\mathbb{P}\{I = I^* \mid \Gamma_{I^*} > \gamma_{I^*}\}. \tag{5.57}$$

If $\mathbb{P}\{I = I^* \mid \Gamma_{I^*} > \gamma_{I^*}\} < 1$, then from (5.57) it follows that $P_{\text{cov}}(I) < P_{\text{cov}}(I^*)$. □

Remark 5.6 A *selection bias* is often introduced in HCNs in order to get UEs to associate preferentially with certain tiers as opposed to others, and thereby obtain a desired distribution of load across the tiers. This is implemented by selecting the serving tier according to (see Section 5.3.2)

$$I = \arg \max_{1 \le i \le n_{\text{open}}} \tau_i Y_{B_i}, \quad \tau_i > 0,\ i = 1, \ldots, n_{\text{open}}, \tag{5.58}$$

instead of (5.49), where Y_{B_i} is the received power at the UE from the candidate serving BS B_i in tier i and τ_i is the selection bias for tier i, $i = 1, \ldots, n_{\text{open}}$. It is clear that serving tier selection with bias does not always yield the max-SINR serving tier I^*, even if we replace the received powers Y_{B_i} from the candidate serving BSs by the received SINRs Γ_i in (5.58) for $i = 1, \ldots, n_{\text{open}}$. In other words, $\mathbb{P}\{I = I^* \mid \Gamma_{I^*} > \gamma_{I^*}\} < 1$, so from (5.56) it follows that biased serving tier selection schemes yield poorer coverage than the max-SINR serving tier selection.

Coverage probability under max-SINR serving BS selection

The joint CCDF (5.38) and the inclusion-exclusion formula applied to (5.48) allow us to write explicit expressions for the coverage probability P_{cov} in general. Corresponding explicit expressions for the coverage probability $P_{\text{cov}}(I^*)$ under the max-SINR rule for serving BS selection are harder to obtain, and require results from Section 5.3. However, for the three cases in Theorem 5.8 where $P_{\text{cov}} = P_{\text{cov}}(I^*)$, $P_{\text{cov}}(I^*)$ can be obtained from the expression for P_{cov}, which we now calculate for these cases. We also assume that $N_0 = 0$ and $\alpha_1 = \cdots = \alpha_{n_{\text{tier}}} = \alpha$, say.

We begin with the simplest expression for P_{cov}: for the third case in Theorem 5.8, it follows from (5.55) that

$$[\gamma_1, \ldots, \gamma_{n_{\text{open}}}]^\top \equiv \boldsymbol{\gamma} > \mathbf{1}_{n_{\text{open}}} \Rightarrow P_{\text{cov}} = \sum_{i=1}^{n_{\text{open}}} \mathbb{P}\{\Gamma_i > \gamma_i\}. \tag{5.59}$$

From (5.44), we then have

$$N_0 = 0,\ \alpha_1 = \cdots = \alpha_{n_{\text{tier}}} = \alpha, \boldsymbol{\gamma} > \mathbf{1}_{n_{\text{open}}} \Rightarrow \mathbb{P}\{\Gamma_{I^*} > \gamma_{I^*}\} = \sum_{i=1}^{n_{\text{open}}} \beta_i \frac{\operatorname{sinc}(2/\alpha)}{\gamma_i^{2/\alpha}}.$$

For the second case in Theorem 5.8, $P_{\text{cov}} = P_{\text{cov}}(I^*)$ is also the marginal CCDF of the SINR at the UE when the serving BS is selected using the max-SINR rule:

$$\boldsymbol{\gamma} = \gamma \mathbf{1}_{n_{\text{open}}} \Rightarrow \mathbb{P}\{\Gamma_{I^*} > \gamma\} = \mathbb{P}\Bigg(\bigcup_{i=1}^{n_{\text{open}}} \{\Gamma_i > \gamma\}\Bigg), \quad \gamma > 0, \quad n_{\text{open}} \geq 1,$$

which also subsumes the first case in Theorem 5.8. From the argument leading to the derivation of (5.43), we know that $\mathbb{P}(\cup_{i=1}^{n_{\text{open}}}\{\Gamma_i > \gamma\})$, the probability that there is at least one BS in one of the open tiers such that the received SINR from this BS exceeds γ, can be expressed as the sum over all $n \geq 1$ of the probability that (a) the SINR at the user when receiving from n BSs of the single-tier deployment Φ^* exceeds γ, and that (b) these n BSs all have "colors" from the set $\{1, \ldots, n_{\text{open}}\}$. The probability of (b) is just $(\sum_{i=1}^{n_{\text{open}}} \beta_i)^n$, while, if $N_0 = 0$ and $\alpha_1 = \cdots = \alpha_{n_{\text{open}}} = \alpha$, say, the probability of (a) is given by (5.42). Thus we have

$$\begin{aligned} N_0 = 0,\ \alpha_1 = \cdots = \alpha_{n_{\text{tier}}} = \alpha \Rightarrow\ & \mathbb{P}\Bigg(\bigcup_{i=1}^{n_{\text{open}}} \{\Gamma_i > \gamma\}\Bigg) \\ = & \sum_{n=1}^{\lceil \rho^{-1} \rceil - 1} \frac{(-1)^{n-1}[\operatorname{sinc}(2/\alpha)]^n}{n(1-n\rho)} \left(\frac{1}{\rho} - n\right)^{2n/\alpha} T_n(\rho, \alpha) \left(\sum_{i=1}^{n_{\text{open}}} \beta_i\right)^n, \quad \gamma > 0. \end{aligned} \tag{5.60}$$

Note that if all tiers are accessible to the UE, i.e. if $n_{\text{open}} = n_{\text{tier}}$, then $\sum_{i=1}^{n_{\text{open}}} \beta_i = \sum_{i=1}^{n_{\text{tier}}} \beta_i = 1$, and then the right side of (5.60) is the same as that of (4.74), which is the probability of coverage in a single-tier deployment. In other words, *deploying additional tiers of accessible BSs on top of a single tier does not affect UE coverage*, at least when the system is interference limited ($N_0 = 0$), the path-loss exponents are the same on all links to the UE, and the coverage threshold for the SINR is the same in all tiers. Thus with additional tiers of BSs we can reap the capacity benefits of the spatial reuse of resources in the cells without adversely affecting UE coverage.

Further, if all tiers are accessible ($n_{\text{open}} = n_{\text{tier}}$), we can calculate $\mathbb{P}\{\Gamma_{I^*} > \gamma_{I^*}\}$, the probability of coverage under the max-SINR rule for serving BS selection, in the case where $\gamma_1, \ldots, \gamma_{n_{\text{tier}}}$ are not necessarily equal, as follows. First, we write

$$\begin{aligned} \mathbb{P}\{\Gamma_{I^*} > \gamma_{I^*}\} &= \mathbb{P}\{\max_{1 \leq j \leq n_{\text{tier}}} \Gamma_j > \gamma_{I^*}\} \\ &= \sum_{i=1}^{n_{\text{tier}}} \mathbb{P}\{I^* = i, \max_{1 \leq j \leq n_{\text{tier}}} \Gamma_j > \gamma_i\} \\ &= \sum_{i=1}^{n_{\text{tier}}} \mathbb{P}\{I^* = i\}\, \mathbb{P}\{\max_{1 \leq j \leq n_{\text{tier}}} \Gamma_j > \gamma_i\}. \end{aligned} \tag{5.61}$$

Next, we observe from the argument leading to the derivation of (5.43) that

$$\mathbb{P}\{\max_{1 \le j \le n_{\text{tier}}} \Gamma_j > \gamma_i\} = \mathbb{P}\{\Gamma^* > \gamma_i\},$$

where Γ^* is the maximum SINR at the UE when receiving from the BSs in the single-tier deployment Φ^*. From (4.74), we therefore have:

$$\begin{aligned} N_0 = 0,\ \alpha_1 = \cdots = \alpha_{n_{\text{tier}}} = \alpha \Rightarrow \mathbb{P}\{\Gamma^* > \gamma_i\} \\ = \sum_{n=1}^{\lceil \rho_i^{-1} \rceil - 1} \frac{(-1)^{n-1}[\text{sinc}(2/\alpha)]^n}{n(1-n\rho_i)} \left(\frac{1}{\rho_i} - n\right)^{2n/\alpha} T_n(\rho_i, \alpha), \\ \rho_i = \frac{\gamma_i}{1+\gamma_i},\ i = 1, \ldots, n_{\text{tier}}. \end{aligned} \tag{5.62}$$

Also, for any $i = 1, \ldots, n_{\text{tier}}$, $\mathbb{P}\{I^* = i\}$ is just the probability that the BS in Φ^* that gives maximum SINR at the UE has "color" i, which is β_i:

$$\mathbb{P}\{I^* = i\} = \beta_i, \quad i = 1, \ldots, n_{\text{tier}}. \tag{5.63}$$

Substituting (5.63) and (5.62) into (5.61), we obtain the coverage probability under the max-SINR rule for serving BS selection with arbitrary coverage thresholds in the tiers, when all tiers are accessible, thermal noise is negligible, and the path-loss exponents are equal in all tiers:

$$\begin{aligned} n_{\text{open}} = n_{\text{tier}}, N_0 = 0, \alpha_1 = \cdots = \alpha_{n_{\text{tier}}} = \alpha \Rightarrow \mathbb{P}\{\Gamma_{I^*} > \gamma_{I^*}\} \\ = \sum_{i=1}^{n_{\text{tier}}} \beta_i \sum_{n=1}^{\lceil \rho_i^{-1} \rceil - 1} \frac{(-1)^{n-1}[\text{sinc}(2/\alpha)]^n}{n(1-n\rho_i)} \left(\frac{1}{\rho_i} - n\right)^{2n/\alpha} T_n(\rho_i, \alpha), \quad \rho_i = \frac{\gamma_i}{1+\gamma_i}. \end{aligned} \tag{5.64}$$

Alternative expressions for (5.63) and (5.64) have been derived in Madhusudhanan *et al.* (2012a).

5.3 Distributions of serving tier and SINR from serving BS

The (single) BS that *actually* serves the user is chosen from among the candidate serving BSs from the accessible tiers according to some criterion. The actual SINR at the user is that on the link from this serving BS to the user. While the selection criterion for the candidate serving BS from each tier can be "nearest" or "strongest" (which is equivalent to maximum SINR at the user), the selection criteria for the actual serving BS can also incorporate a *selection bias* in order to bias the selection of the serving BS toward candidate serving BSs from certain tiers as opposed to others. When applied to favor selection of a femtocell in a macro-femto HCN, for example, this has the benefit of achieving greater offloading of traffic from the macrocellular network.

Each tier has a bias $\tau_i > 0$, $i = 1, \ldots, n_{\text{open}}$, where the meaning and dimension of $\{\tau_i\}_{i=1}^{n_{\text{open}}}$ are different for each serving BS selection criterion (see below). For future use, we define the n_{open} vectors $\boldsymbol{\theta}^{(i)}$, each with n_{open} entries, where the entries in $\boldsymbol{\theta}^{(i)}$ are dimensionless for each $i = 1, \ldots, n_{\text{open}}$:

$$\boldsymbol{\theta}^{(i)} \equiv \left[\theta_1^{(i)}, \ldots, \theta_{n_{\text{open}}}^{(i)}\right]^{\mathsf{T}} = \left[\frac{\tau_i}{\tau_1}, \ldots, \frac{\tau_i}{\tau_{n_{\text{open}}}}\right]^{\mathsf{T}} \in \mathbb{R}^{n_{\text{open}}}, \quad i = 1, \ldots, n_{\text{open}}.$$

Recall that the tier that actually serves the user is denoted by the random variable I, taking values in $\{1, \ldots, n_{\text{open}}\}$. The actual serving BS is the candidate serving BS from tier I, and the actual SINR Γ at the user is that when receiving from this BS:

$$\Gamma \equiv \Gamma_I.$$

In this section, we derive both the marginal SINR distribution $\mathbb{P}\{\Gamma > \gamma\}$ and the PMF of the serving tier $\mathbb{P}\{I = i\}$ under different selection criteria for the candidate serving BSs in the tiers, the overall serving BS, and selection biases for the tiers.

5.3.1 Serving BS is the "nearest" (after selection bias) candidate serving BS

The serving BS is selected as the candidate serving BS from tier

$$I = \arg \min_{1 \le i \le n_{\text{open}}} \tau_i R_i^{\alpha_i},$$

where R_i is the distance of the candidate serving BS from tier i, $i = 1, \ldots, n_{\text{open}}$. This serving BS selection criterion only makes sense if coupled with the nearest-BS rule for the selection of the candidate serving BSs in the open tiers.

Distribution of serving tier

This serving BS selection criterion is equivalent to selecting the serving BS based on strongest *averaged* (rather than *instantaneous*) power at the user after accounting for selection bias (Jo *et al.*, 2012). It is easy to see that

$$\mathbb{P}\{I = i\} = \mathbb{P}\left\{R_j > \left[\theta_j^{(i)} R_i^{\alpha_i}\right]^{1/\alpha_j}, \, j \in \{1, \ldots, n_{\text{open}}\} \setminus \{i\}\right\}, \quad i = 1, \ldots, n_{\text{open}}. \tag{5.65}$$

Note that (see Problem 4.11) for any x,

$$\mathbb{P}\{R_j > x\} = \exp\left(-\pi \lambda_j x^2\right), \; x \ge 0, \quad j = 1, \ldots, n_{\text{open}}.$$

Then, conditioning on $R_i = r_i$ and using the independence of the BS location PPPs across the tiers, we obtain

$$\begin{aligned} &\mathbb{P}\left\{R_j > \left[\theta_j^{(i)} R_i^{\alpha_i}\right]^{1/\alpha_j}, \, j \in \{1, \ldots, n_{\text{open}}\} \setminus \{i\} \;\middle|\; R_i = r_i\right\} \\ &= \prod_{\substack{j=1 \\ j \ne i}}^{n_{\text{open}}} \mathbb{P}\left\{R_j > \left[\theta_j^{(i)} r_i^{\alpha_i}\right]^{1/\alpha_j}\right\} = \exp\left\{-\pi \sum_{\substack{j=1 \\ j \ne i}}^{n_{\text{open}}} \lambda_j \left[\theta_j^{(i)} r_i^{\alpha_i}\right]^{2/\alpha_j}\right\}. \end{aligned} \tag{5.66}$$

Substituting (5.66) into the right side of (5.65), integrating with respect to the PDF $f_{R_i}(\cdot)$ of R_i given by $f_{R_i}(r_i) = 2\pi\lambda_i r_i \exp(-\pi\lambda_i r_i^2)$, and making the change of variables from r_i to $u_i = \pi\lambda_i r_i^2$, we obtain the PMF of the serving tier I as follows:

$$\begin{aligned}
\mathbb{P}\{I = i\} &= \int_0^\infty f_{R_i}(r_i)\, \mathbb{P}\left\{R_j > \left[\theta_j^{(i)} R_i^{\alpha_i}\right]^{1/\alpha_j},\, j \in \{1, \ldots, n_{\text{open}}\} \setminus \{i\} \mid R_i = r_i\right\} \mathrm{d}r_i \\
&= \int_0^\infty \exp\left\{-u_i - \pi \sum_{\substack{j=1 \\ j \neq l}}^{n_{\text{open}}} \lambda_j \left[\theta_j^{(i)} \left(\frac{u_i}{\pi\lambda_i}\right)^{\alpha_i/2}\right]^{2/\alpha_j}\right\} \mathrm{d}u_i \\
&= \int_0^\infty \exp\left\{-\pi \sum_{j=1}^{n_{\text{open}}} \lambda_j \left[\theta_j^{(i)} \left(\frac{u_i}{\pi\lambda_i}\right)^{\alpha_i/2}\right]^{2/\alpha_j}\right\} \mathrm{d}u_i, \quad i = 1, \ldots, n_{\text{open}}.
\end{aligned} \tag{5.67}$$

The distribution of the serving tier (5.67) was proved in Jo *et al.* (2012).

Distribution of SINR from the serving BS

The distribution of the SINR Γ_I is given by the following theorem, which provides the marginal SINR distribution result to complement the joint conditional SINR distribution result in Theorem 5.2.

THEOREM 5.10 *If the candidate serving BSs from the open tiers are those nearest to the user, the fade attenuation on the links from the candidate serving BS in tier i has PDF $f_{H_{i,*}}(\cdot)$, $i = 1, \ldots, n_{\text{open}}$, and the serving BS is the candidate serving BS from the serving tier $I = \arg\min_{1 \le i \le n_{\text{open}}} \tau_i R_i^{\alpha_i}$, then the distribution of the SINR at the user when receiving from this serving BS is given by*

$$\begin{aligned}
&\mathbb{P}\{\Gamma_I > \gamma\} \\
&= \sum_{i=1}^{n_{\text{open}}} \int_0^\infty \mathbb{P}\left\{\Gamma_i > \gamma \mid R_i = r_i,\, R_j > \left[\theta_j^{(i)} r_i^{\alpha_i}\right]^{1/\alpha_j},\, j \in \{1, \ldots, n_{\text{open}}\} \setminus \{i\}\right\} \\
&\quad \times 2\pi\lambda_i r_i \exp\left\{-\pi \sum_{j=1}^{n_{\text{open}}} \lambda_j \left[\theta_j^{(i)} r_i^{\alpha_i}\right]^{2/\alpha_j}\right\} \mathrm{d}r_i, \quad \gamma > 0,
\end{aligned}$$

where, for any $i = 1, \ldots, n_{\text{open}}$,

$$\begin{aligned}
&\mathbb{P}\left\{\Gamma_i > \gamma \mid R_i = r_i,\, R_j > \left[\theta_j^{(i)} r_i^{\alpha_i}\right]^{1/\alpha_j},\, j \in \{1, \ldots, n_{\text{open}}\} \setminus \{i\}\right\} \\
&\qquad = \sum_{l=1}^{\infty} \left\{\frac{(-1)^{m_{i,l}} \beta_{i,l} c_{i,l}^{m_{i,l}+1}}{m_{i,l}!}\right\} \frac{\partial^{m_{i,l}}}{\partial c_{i,l}^{m_{i,l}}} c_{i,l}^{-1} \mathcal{L}_{W \mid R_i = r_i}\left(\frac{c_{i,l} r_i^{\alpha} \gamma}{P_i}\right),
\end{aligned}$$

when PDF $f_{H_{i,*}}(\cdot)$ *is expressed as an infinite mixture of Erlang PDFs in the form* (5.9) *and*

$$\mathcal{L}_{W \mid R_i=r_i}(s) = \exp\left\{-sN_0 - \pi \sum_{j=1}^{n_{\text{open}}} \lambda_j \left(sP_j\right)^{2/\alpha_j} G_{j,\alpha_j}\left[\left(\frac{\theta_j^{(i)} r_i^{\alpha_i}}{P_j s}\right)^{2/\alpha_j}\right] - \pi \sum_{k=n_{\text{open}}+1}^{n_{\text{tier}}} \lambda_k \left(sP_k\right)^{2/\alpha_k} G_{k,\alpha_k}(0)\right\}, \quad s > 0.$$

Proof

$$\begin{aligned}
&\mathbb{P}\{\Gamma_I > \gamma\} \\
&= \sum_{i=1}^{n_{\text{open}}} \mathbb{P}\{\Gamma_i > \gamma,\, I = i\} \\
&= \sum_{i=1}^{n_{\text{open}}} \mathbb{P}\left\{\Gamma_i > \gamma,\, R_j^{\alpha_j} > \theta_j^{(i)} R_i^{\alpha_i},\, j \in \{1,\ldots,n_{\text{open}}\} \setminus \{i\}\right\} \\
&= \sum_{i=1}^{n_{\text{open}}} \int_0^\infty \mathbb{P}\left\{\Gamma_i > \gamma,\, R_j^{\alpha_j} > \theta_j^{(i)} r_i^{\alpha_i},\, j \in \{1,\ldots,n_{\text{open}}\} \setminus \{i\} \mid R_i = r_i\right\} f_{R_i}(r_i)\mathrm{d}r_i \\
&= \sum_{i=1}^{n_{\text{open}}} \int_0^\infty \mathbb{P}\left\{\Gamma_i > \gamma \mid R_i = r_i,\, R_j > \left[\theta_j^{(i)} r_i^{\alpha_i}\right]^{1/\alpha_j},\, j \in \{1,\ldots,n_{\text{open}}\} \setminus \{i\}\right\} \\
&\quad \times \mathbb{P}\left\{R_j > \left[\theta_j^{(i)} r_i^{\alpha_i}\right]^{1/\alpha_j},\, j \in \{1,\ldots,n_{\text{open}}\} \setminus \{i\} \mid R_i = r_i\right\} f_{R_i}(r_i)\mathrm{d}r_i \\
&= \sum_{i=1}^{n_{\text{open}}} \int_0^\infty \mathbb{P}\left\{\Gamma_i > \gamma \mid R_i = r_i,\, R_j > \left[\theta_j^{(i)} r_i^{\alpha_i}\right]^{1/\alpha_j},\, j \in \{1,\ldots,n_{\text{open}}\} \setminus \{i\}\right\} \\
&\quad \times 2\pi\lambda_i r_i \exp\left\{-\pi \sum_{j=1}^{n_{\text{open}}} \lambda_j \left[\theta_j^{(i)} r_i^{\alpha_i}\right]^{2/\alpha_j}\right\} \mathrm{d}r_i, \qquad (5.68)
\end{aligned}$$

where in the final step we use the independence of the tier PPPs, the PDFs of $R_1, \ldots, R_{n_{\text{open}}}$, and the fact that $\theta_i^{(i)} \equiv 1$. From Remark 5.2, it follows that

$$\mathbb{P}\left\{\Gamma_i > \gamma \mid R_i = r_i,\, R_j > \left[\theta_j^{(i)} r_i^{\alpha_i}\right]^{1/\alpha_j},\, j \in \{1,\ldots,n_{\text{open}}\} \setminus \{i\}\right\}$$

is given by (5.8) with $k = 1$ and $\Gamma_1, \gamma_1, R_1, r_1$ replaced by $\Gamma_i, \gamma, R_i, r_i$, respectively, and $\mathcal{L}_{W \mid R_i=r_i}(s)$ is given by (5.18) with

$$d_{\min,j} = \begin{cases} \left[\theta_j^{(i)} r_i^{\alpha_i}\right]^{1/\alpha_j}, & j \in \{1,\ldots,n_{\text{open}}\} \setminus \{i\}, \\ 0, & j = n_{\text{open}}+1,\ldots,n_{\text{tier}}. \end{cases}$$

Substituting into (5.68) completes the proof. □

5.3.2 Serving BS is the "strongest" (after selection bias) candidate serving BS

Using the notation of Section 2.1.2, the serving BS is the candidate serving BS B_I from tier $I = \arg\max_{1 \le i \le n_{\text{open}}} \tau_i Y_{B_i}$, where B_i is the candidate serving BS from tier i, $i = 1, \ldots, n_{\text{open}}$. We are interested in the PMF of the serving tier,

$$\mathbb{P}\{I = i\}, \quad i = 1, \ldots, n_{\text{open}}, \tag{5.69}$$

and in the marginal CCDF of the SINR Γ_I at the user when receiving from the serving BS,

$$\mathbb{P}\{\Gamma_I > \gamma\} = \sum_{i=1}^{n_{\text{open}}} \mathbb{P}\{I = i, \Gamma_i > \gamma\}, \quad \gamma > 0. \tag{5.70}$$

Selection from pilot power, SINR from traffic channel power

Strictly speaking, in cellular networks, serving BS selection is usually made on the basis of received power on the *pilot* channels, so $I = \arg\max_{1 \le i \le n_{\text{open}}} \tau_i Y_{B_i}^{\text{pilot}}$, where

$$Y_b^{\text{pilot}} = \frac{K_i P_i^{\text{pilot,tx}} H_{i,b}^{\text{pilot}}}{R_b^{\alpha_i}}, \quad b \in \Phi_i, \quad i = 1, \ldots, n_{\text{open}},$$

and $P_i^{\text{pilot,tx}}$ is the transmit power of the BSs belonging to tier i on the pilot channel, $i = 1, \ldots, n_{\text{open}}$. However, we are interested in the SINR at the user on the *traffic* channel, which is transmitted with a different power (also assumed the same across all BSs in a tier) $P_i^{\text{traffic,tx}}$ by the BSs in tier i, $i = 1, \ldots, n_{\text{open}}$. We assume all BSs in closed tiers $j \in \{n_{\text{open}} + 1, \ldots, n_{\text{tier}}\}$ transmit on the traffic channel with power $P_j^{\text{traffic,tx}}$, $j = n_{\text{open}} + 1, \ldots, n_{\text{tier}}$. Then the received power at the user from BS $b \in \Phi_i$ is given by

$$Y_b^{\text{traffic}} = \frac{K_i P_i^{\text{traffic,tx}} H_{i,b}^{\text{traffic}}}{R_b^{\alpha_i}}, \quad b \in \Phi_i, \quad i = 1, \ldots, n_{\text{tier}},$$

where we have written the fade attenuation on the traffic channel as $H_{i,b}^{\text{traffic}}$ to distinguish it from $H_{i,b}^{\text{pilot}}$, the fade attenuation on the pilot channel. We assume for simplicity that $H_{i,b}^{\text{traffic}}$ and $H_{i,b}^{\text{pilot}}$ have the same distribution.

The SINR on the traffic channel when the user receives from the candidate serving BS in tier i is given by

$$\Gamma_i^{\text{traffic}} = \frac{Y_{B_i}^{\text{traffic}}}{\sum_{b \in \Phi_i \setminus \{B_i\}} Y_b^{\text{traffic}} + \sum_{\substack{j=1 \\ j \ne i}}^{n_{\text{tier}}} \sum_{b \in \Phi_j} Y_b^{\text{traffic}} + N_0}, \quad i = 1, \ldots, n_{\text{open}},$$

and (5.70) becomes

$$\mathbb{P}\{\Gamma_I^{\text{traffic}} > \gamma\} = \sum_{i=1}^{n_{\text{open}}} \mathbb{P}\left\{i = \arg\max_{1 \le j \le n_{\text{open}}} \tau_j Y_{B_j}^{\text{pilot}}, \Gamma_i^{\text{traffic}} > \gamma\right\}, \quad \gamma > 0. \tag{5.71}$$

Consider the idealized model where

(1) the ratio $P_i^{\text{traffic,tx}}/P_i^{\text{pilot,tx}}$ of the transmit power on the traffic channel to the transmit power on the pilot channel is the same for all $i = 1, \ldots, n_{\text{open}}$, and
(2) $H_{i,b}^{\text{traffic}} \equiv H_{i,b}^{\text{pilot}} \equiv H_{i,b}$, say, for all $b \in \Phi_i$ and all $i = 1, \ldots, n_{\text{open}}$, i.e. serving tier selection (based on pilot powers) and transmission from the selected serving BS occur close enough in time that the fade attenuation may be assumed unchanged.

Under these assumptions, we see that we may simplify (5.71) to (5.70), where all quantities are that for the *traffic* channel alone. This is the assumption we make for the remainder of this section. Then we can also replace the pilot channel quantities in (5.69) with the corresponding traffic channel quantities.

In short, for the remainder of this section, we assume that all quantities in (5.69) and (5.70) are those for the traffic channel. For simplicity of notation, we drop the superscript, e.g. we write Γ_I instead of $\Gamma_I^{\text{traffic}}$, etc.

Serving tier distribution from SINR distribution

The key calculation is that of

$$\mathbb{P}\{I = i, \Gamma_i > \gamma\}, \quad \gamma > 0, \quad i = 1, \ldots, n_{\text{open}}. \tag{5.72}$$

The analysis is different for "nearest" and "strongest" criteria for selecting the candidate serving BSs B_i, $i \in \{1, \ldots, n_{\text{open}}\}$. We assume that the same criterion (either "nearest" or "strongest") is applied for candidate serving BS selection in every tier.

Remark 5.7 It is easy to see that

$$\mathbb{P}\{I = i\} = \lim_{\gamma \downarrow 0} \mathbb{P}\{I = i, \Gamma_i > \gamma\}, \quad i = 1, \ldots, n_{\text{open}},$$

so in theory it is sufficient to calculate (5.72) in order to obtain the PMF of the serving tier. Unfortunately, the expressions we shall derive for (5.72) are not in closed form and require numerical evaluation. Moreover, as γ becomes very small, these expressions become numerically unstable. It is thus better in practice to derive expressions for (5.69) directly.

SINR distribution when candidate serving BS in each tier is the nearest to the user

Define $\boldsymbol{R} = [R_1, \ldots, R_{n_{\text{open}}}]^\top$ to be the vector of distances to the user from the candidate serving BSs in tiers $1, \ldots, n_{\text{open}}$, respectively. We make our usual assumptions that all fades on all links to the user are independent, and that the fade attenuations on the links to the user from the BSs in tier j, $j = 1, \ldots, n_{\text{tier}}$, are identically distributed with arbitrary PDF $f_{H_j}(\cdot)$, except possibly for the fade attenuation $H_{i,*}$ on the link from the candidate serving BS in tier i, *provided it is the actual serving BS*, $i = 1, \ldots, n_{\text{open}}$. In other words, only the serving BS may have a fade attenuation to the user with a distribution different from that of any other BS in the same tier. We begin by noting that in (5.70)

$$\begin{aligned}\mathbb{P}\{I = i, \Gamma_i > \gamma\} \\ &= \mathbb{P}\left\{\Gamma_i > \gamma,\ \frac{P_j H_j}{R_j^{\alpha_j}} \le \theta_j^{(i)} \frac{P_i H_{i,*}}{R_i^{\alpha_i}},\ j \in \{1, \ldots, n_{\text{open}}\} \setminus \{i\}\right\} \\ &= \mathbb{P}\{\Gamma_i > \gamma\} - \mathbb{P}\left(\{\Gamma_i > \gamma\} \cap \bigcup_{\substack{j=1 \\ j \ne i}}^{n_{\text{open}}} \left\{\frac{P_j H_j}{R_j^{\alpha_j}} > \theta_j^{(i)} \frac{P_i H_{i,*}}{R_i^{\alpha_i}}\right\}\right), \quad i = 1, \ldots, n_{\text{open}}.\end{aligned} \tag{5.73}$$

For each $i = 1, \ldots, n_{\text{open}}$, $\mathbb{P}\{\Gamma_i > \gamma\}$ in (5.73) can be calculated from

$$\begin{aligned}\mathbb{P}\{\Gamma_i > \gamma\} &= \int_0^\infty \mathbb{P}\{\Gamma_i > \gamma \mid R_i = r_i\} f_{R_i}(r_i)\, \mathrm{d}r_i, \\ f_{R_i}(r_i) &= 2\pi\lambda_i r_i \exp(-\pi\lambda_i r_i^2),\ r_i \ge 0,\end{aligned} \tag{5.74}$$

where $\mathbb{P}\{\Gamma_i > \gamma \mid R_i = r_i\}$ is calculated from (5.8) with $k = 1$ and $\Gamma_1, \gamma_1, R_1, r_1$ replaced by $\Gamma_i, \gamma, R_i, r_i$, respectively.

Now, for each $i \in \{1, \ldots, n_{\text{open}}\}$, the second term in the summation in (5.73) can be evaluated using the inclusion-exclusion formula:

$$\begin{aligned}&\mathbb{P}\left(\{\Gamma_i > \gamma\} \cap \bigcup_{\substack{j=1 \\ j \ne i}}^{n_{\text{open}}} \left\{\frac{P_j H_j}{R_j^{\alpha_j}} > \theta_j^{(i)} \frac{P_i H_{i,*}}{R_i^{\alpha_i}}\right\}\right) \\ &= \sum_{k=2}^{n_{\text{open}}} (-1)^k \sum_{\substack{i_2, \ldots, i_k: \\ 1 \le i_2 < \cdots < i_k \le n_{\text{open}} \\ i_l \ne i, l = 2, \ldots, k}} \mathbb{P}\left(\{\Gamma_i > \gamma\} \cap \bigcap_{l=2}^{k} \left\{\frac{P_{i_l} H_{i_l}}{R_{i_l}^{\alpha_{i_l}}} > \theta_{i_l}^{(i)} \frac{P_i H_{i,*}}{R_i^{\alpha_i}}\right\}\right).\end{aligned} \tag{5.75}$$

Let us now focus on a single term in the multiple-summation on the right side of (5.75). Following the same steps as in the derivation of (2.10), we can show (see Problem 5.4) that, conditioned on $R_1 = r_1, \ldots, R_k = r_k$, for any $k = 2, \ldots, n_{\text{open}}$ and any $\theta_2 > 0, \ldots, \theta_k > 0$, the set of relationships

$$\Gamma_1 > \gamma,\ \frac{P_2 H_2}{r_2^{\alpha_2}} > \theta_2 \frac{P_1 H_{1,*}}{r_1^{\alpha_1}},\ \ldots,\ \frac{P_k H_k}{r_k^{\alpha_k}} > \theta_k \frac{P_1 H_{1,*}}{r_1^{\alpha_1}}$$

is equivalent to the set of relationships

$$\mathbf{A}_k(\boldsymbol{\theta})\boldsymbol{X} > W\boldsymbol{e}_1^{(k)}, \tag{5.76}$$

conditioned on $R_1 = r_1, \ldots, R_k = r_k$, where

$$\begin{aligned}
\boldsymbol{\theta} &= [\theta_1, \theta_2, \ldots, \theta_k]^\top, \quad \theta_1 \equiv 1, \\
\mathbf{A}_k(\boldsymbol{\theta}) &= \mathbf{I}_k + [\rho^{-1}\boldsymbol{e}_1^{(k)} - \boldsymbol{\theta}]\big(\boldsymbol{e}_1^{(k)}\big)^\top - \boldsymbol{e}_1^{(k)}\mathbf{1}_k^\top, \quad \rho \equiv \gamma/(1+\gamma), \\
\boldsymbol{X} &= [X_1, \ldots, X_k]^\top, \quad X_1 = P_1 H_{1,*}/r_1^{\alpha_1}, \; X_j = P_j H_j / r_j^{\alpha_j}, \; j = 2, \ldots, k, \\
W &= \sum_{i=1}^{k} \sum_{b \in \Phi_i \setminus \{B_i\}} Y_b + \sum_{j=k+1}^{n_{\text{tier}}} \sum_{b \in \Phi_j} Y_b + N_0.
\end{aligned}$$

In other words, each of the probabilities on the right side of (5.75) can be expressed, by relabeling the tiers if necessary, in the form

$$\begin{aligned}
&\mathbb{P}\left(\{\Gamma_1 > \gamma\} \cap \bigcap_{j=2}^{k} \left\{\frac{P_j H_j}{R_j^{\alpha_j}} > \theta_j \frac{P_1 H_{1,*}}{R_1^{\alpha_1}}\right\} \,\middle|\, R_1 = r_1, \ldots, R_k = r_k\right) \\
&= \mathbb{P}\left\{\Gamma_1 > \gamma, \; \frac{P_2 H_2}{r_2^{\alpha_2}} > \theta_2 \frac{P_1 H_{1,*}}{r_1^{\alpha_1}}, \ldots, \frac{P_k H_k}{r_k^{\alpha_k}} > \theta_k \frac{P_1 H_{1,*}}{r_1^{\alpha_1}} \,\middle|\, R_1 = r_1, \ldots, R_k = r_k\right\} \\
&= \mathbb{P}\{\mathbf{A}_k(\boldsymbol{\theta})\boldsymbol{X} > W\boldsymbol{e}_1^{(k)} \mid R_1 = r_1, \ldots, R_k = r_k\}
\end{aligned}$$

for some $\boldsymbol{\theta}$, $\boldsymbol{X}$, and W as just defined. It is easy to show (see Problem 5.5) that $\mathbf{A}_k(\boldsymbol{\theta})$ is a Z-matrix satisfying (2.25), and that

$$\det \mathbf{A}_k(\boldsymbol{\theta}) = \mathbf{1}_k^\top \mathbf{A}_k(\boldsymbol{\theta})\boldsymbol{e}_1^{(k)} = \rho^{-1} - \mathbf{1}_k^\top \boldsymbol{\theta}. \tag{5.77}$$

Further, we can verify directly that

$$\begin{aligned}
\mathbf{A}_k(\boldsymbol{\theta})^{-1} &= \mathbf{I}_k + \frac{\boldsymbol{\theta}\mathbf{1}_k^\top}{\frac{1}{\rho} - \mathbf{1}_k^\top \boldsymbol{\theta}} - \boldsymbol{e}_1^{(k)}\big(\boldsymbol{e}_1^{(k)}\big)^\top, \\
\mathbf{A}_k(\boldsymbol{\theta})^{-1}\boldsymbol{e}_i^{(k)} &= (1 - \delta_{1,i})\boldsymbol{e}_i^{(k)} + \frac{\boldsymbol{\theta}}{\frac{1}{\rho} - \mathbf{1}_k^\top \boldsymbol{\theta}}, \quad i = 1, \ldots, k.
\end{aligned} \tag{5.78}$$

Next, we use the transformation (5.11) to write

$$\mathbf{A}_k(\boldsymbol{\theta})\boldsymbol{X} > W\boldsymbol{e}_1^{(k)} \Leftrightarrow \mathbf{A}_{k,*}(\boldsymbol{\theta}_*)\boldsymbol{H}_* > W\mu_{1,*}^{-1}\boldsymbol{e}_1^{(k)}, \quad \boldsymbol{H}_* \equiv [H_{1,*}, H_2, \ldots, H_k]^\top,$$

where $\boldsymbol{\mu}_*$ is defined in (5.10), and

$$\begin{aligned}
\mathbf{A}_{k,*}(\boldsymbol{\theta}_*) &= \mathbf{M}_*^{-1}\mathbf{A}_k(\boldsymbol{\theta})\mathbf{M}_* = \mathbf{I}_k + \left(\rho_*^{-1}\boldsymbol{e}_1^{(k)} - \boldsymbol{\theta}_*\right)\boldsymbol{e}_1^{(k)} - \mu_{1,*}^{-1}\boldsymbol{e}_1^{(k)}\boldsymbol{\mu}_*^\top, \\
\boldsymbol{\theta}_* &= \mathbf{M}_*^{-1}\boldsymbol{\theta}\mu_{1,*} = \left[1, \theta_2\frac{\mu_{1,*}}{\mu_{2,*}}, \ldots, \theta_k\frac{\mu_{1,*}}{\mu_{k,*}}\right]^\top \equiv [\theta_{1,*}, \theta_{2,*}, \ldots, \theta_{k,*}]^\top, \\
\rho_* &= \frac{\rho}{\mu_{1,*}}.
\end{aligned}$$

Then we have

$$\det \mathbf{A}_{k,*}(\boldsymbol{\theta}_*) = \det \mathbf{A}_k(\boldsymbol{\theta}) = \rho^{-1} - \mathbf{1}_k^\top \boldsymbol{\theta}$$

and

$$\mathbf{A}_{k,*}(\boldsymbol{\theta}_*)^{-1} = \mathbf{M}_*^{-1}\mathbf{A}_k(\boldsymbol{\theta})^{-1}\mathbf{M}_* = \mathbf{I}_k + \frac{\boldsymbol{\theta}_*\boldsymbol{\mu}_*^\top}{\rho_*^{-1} - \boldsymbol{\mu}_*^\top\boldsymbol{\theta}_*} - \boldsymbol{e}_1^{(k)}\left(\boldsymbol{e}_1^{(k)}\right)^\top,$$

so

$$\mathbf{A}_{k,*}(\boldsymbol{\theta}_*)^{-1}\boldsymbol{e}_i^{(k)} = (1-\delta_{1,i})\boldsymbol{e}_i^{(k)} + \frac{\mu_{i,*}}{\mu_{1,*}}\frac{\boldsymbol{\theta}_*}{\left(\rho^{-1} - \mathbf{1}_k^\top\boldsymbol{\theta}\right)}, \quad i = 1,\ldots,k.$$

It follows from Theorem 2.5 that, if the PDFs $f_{H_{1,*}}(\cdot), f_{H_2}(\cdot), \ldots, f_{H_k}(\cdot)$ are expressed as infinite mixtures of Erlang PDFs in the form (5.9), then each probability on the right side of (5.75) can be evaluated using the following result:

$$\begin{aligned}
&\mathbb{P}\left\{\Gamma_1 > \gamma,\ \frac{P_2H_2}{r_2^{\alpha_2}} > \frac{\theta_2P_1H_{1,*}}{r_1^{\alpha_1}}, \ldots, \frac{P_kH_k}{r_k^{\alpha_k}} > \frac{\theta_kP_1H_{1,*}}{r_1^{\alpha_1}}\,\Big|\, R_1 = r_1, \ldots, R_k = r_k\right\} \\
&\quad = \mathbb{P}\left\{\mathbf{A}_k(\boldsymbol{\theta})\boldsymbol{X} > W\boldsymbol{e}_1^{(k)}\right\} = \mathbb{P}\left\{\mathbf{A}_{k,*}(\boldsymbol{\theta}_*)\boldsymbol{H}_* > W\mu_{1,*}^{-1}\boldsymbol{e}_1^{(k)}\right\} \\
&\quad = \begin{cases} 0, & \gamma \geq (\theta_2 + \cdots + \theta_k)^{-1}, \\ \displaystyle\sum_{l_1=1}^{\infty}\cdots\sum_{l_k=1}^{\infty}\left\{\prod_{j=1}^{k}\frac{(-1)^{m_{j,l_j}}\beta_{j,l_j}(c_{j,l_j})^{m_{j,l_j}+1}}{m_{j,l_j}!}\right\} & \\ \displaystyle\times\frac{\partial^{m_{1,l_1}+\cdots+m_{k,l_k}}}{\partial(c_{1,l_1})^{m_{1,l_1}}\cdots\partial(c_{k,l_k})^{m_{k,l_k}}} & \\ \displaystyle\frac{\mathcal{L}_{W\,|\,R_1=r_1,\ldots,R_k=r_k}\left(\frac{\sum_{n=1}^k c_{n,l_n}\theta_n r_n^{\alpha_n}/P_n}{\rho^{-1} - \mathbf{1}_k^\top\boldsymbol{\theta}}\right)}{\prod_{i=1}^{k}\left[(1-\delta_{1,i})c_{i,l_i} + \mu_{i,*}\frac{\sum_{n=1}^k c_{n,l_n}\theta_n r_n^{\alpha_n}/P_n}{\rho^{-1} - \mathbf{1}_k^\top\boldsymbol{\theta}}\right]}, & \gamma < (\theta_2 + \cdots + \theta_k)^{-1}, \end{cases}
\end{aligned} \tag{5.79}$$

where $\mathcal{L}_{W\,|\,R_1=r_1,\ldots,R_k=r_k}(s)$ is given by (5.5). The corresponding unconditional probability is obtained by integrating with respect to the joint PDF of $R_1, \ldots, R_k$:

$$\begin{aligned}
&\mathbb{P}\left(\{\Gamma_1 > \gamma\} \cap \bigcap_{j=2}^{k}\left\{\frac{P_jH_j}{R_j^{\alpha_j}} > \theta_j\frac{P_1H_{1,*}}{R_1^{\alpha_1}}\right\}\right) = \int_{\mathbb{R}_+^k} \mathrm{d}\boldsymbol{r}\, f_{\boldsymbol{R}}(\boldsymbol{r}) \\
&\times\mathbb{P}\left\{\Gamma_1 > \gamma,\ \frac{P_2H_2}{r_2^{\alpha_2}} > \frac{\theta_2P_1H_{1,*}}{r_1^{\alpha_1}}, \ldots, \frac{P_kH_k}{r_k^{\alpha_k}} > \frac{\theta_kP_1H_{1,*}}{r_1^{\alpha_1}}\,\Big|\, R_1 = r_1, \ldots, R_k = r_k\right\},
\end{aligned} \tag{5.80}$$

where $f_{\boldsymbol{R}}(\boldsymbol{r})$ is given by (5.13).

Collecting all the pieces together, we finally obtain the following theorem.

THEOREM 5.11 *(Mukherjee, 2012b.) If the candidate serving BSs from the open tiers are those nearest to the user, the fade attenuation on the links from the candidate serving BS in tier i has PDF* $f_{H_i}(\cdot)$, $i = 1, \ldots, n_{\text{open}}$, *and the serving BS is the candidate serving BS from the serving tier* $I = \arg\max_{1\leq i\leq n_{\text{open}}} \tau_i Y_{B_i}$, *then the CCDF of the SINR at the user when receiving from this serving BS, conditioned on the distances to the candidate serving BSs in all open tiers, is given from* (5.70) *as the sum of* n_{open} *terms.*

For each $i = 1, \ldots, n_{\text{open}}$*, the corresponding term in* (5.70) *is given from* (5.73) *as the difference of two terms. The first of these two terms is evaluated from* (5.74)*, while the second of these two terms is given by the multiple-summation on the right side of* (5.75)*, where each term in this multiple-summation is in turn calculated from* (5.80).

Remark 5.8 Since the joint CCDF of the SINRs at the user when receiving from the candidate serving BSs in the tiers is a function of only N_0 and $\lambda_i^{\alpha_i/2} K_i P_i^{\text{tx}}$, $i = 1, \ldots, n_{\text{tier}}$, it is not surprising that this is also true of the SINR from the serving BS. This can also be directly verified from (5.79) and (5.80).

Remark 5.9 In the context of (5.71), suppose we assume that, for every BS b in every open tier $i \in \{1, \ldots, n_{\text{open}}\}$, the fade attenuations $H_{i,b}^{\text{traffic}}$ and $H_{i,b}^{\text{pilot}}$ on the links to the user on the pilot and traffic channels are not only identically distributed, but also independent instead of identical (say because the serving tier selection and transmission from the selected serving BS are separated by enough time that the fade attenuations may be assumed independent). Then it is possible to do the analysis directly on (5.71) by conditioning on the distances of the candidate serving BSs in the tiers from the user:

$$\mathbb{P}\left\{i = \arg \max_{1 \leq j \leq n_{\text{open}}} \tau_j Y_{B_j}^{\text{pilot}}, \Gamma_i^{\text{traffic}} > \gamma \mid \boldsymbol{R} = \boldsymbol{r}\right\}$$
$$= \mathbb{P}\left\{i = \arg \max_{1 \leq j \leq n_{\text{open}}} \tau_j Y_{B_j}^{\text{pilot}} \mid \boldsymbol{R} = \boldsymbol{r}\right\} \mathbb{P}\left\{\Gamma_i^{\text{traffic}} > \gamma \mid \boldsymbol{R} = \boldsymbol{r}\right\},$$

and each term on the right can be calculated separately using our existing results and methods. We do not pursue this any further here, but it is an important case for analysis with power control (see Section 6.4.6).

Serving tier distribution when candidate serving BS in each tier is the nearest to the user

It is easy to see that the probability that the serving tier is i is given by

$$\mathbb{P}\{I = i\} = 1 - \mathbb{P}\left(\bigcup_{\substack{j=1 \\ j \neq i}}^{n_{\text{open}}} \left\{\frac{P_j H_j}{R_j^{\alpha_j}} > \theta_j^{(i)} \frac{P_i H_{i,*}}{R_i^{\alpha_i}}\right\}\right)$$
$$= 1 + \sum_{k=2}^{n_{\text{open}}} (-1)^{k-1} \sum_{\substack{i_2, \ldots, i_k: \\ 1 \leq i_2 < \cdots < i_k \leq n_{\text{open}} \\ i_l \neq i, l = 2, \ldots, k}} \mathbb{P}\left(\bigcap_{l=2}^{k} \left\{\frac{P_{i_l} H_{i_l}}{R_{i_l}^{\alpha_{i_l}}} > \theta_{i_l}^{(i)} \frac{P_i H_{i,*}}{R_i^{\alpha_i}}\right\}\right). \quad (5.81)$$

Now, conditioned on $R_1 = r_1, \ldots, R_k = r_k$ for any $k \leq n_{\text{open}}$ and any $\theta_2 > 0, \ldots, \theta_k > 0$,

$$
\begin{aligned}
&\mathbb{P}\left(\bigcap_{j=2}^{k}\left\{\frac{P_j H_j}{R_j^{\alpha_j}} > \theta_j \frac{P_1 H_{1,*}}{R_1^{\alpha_1}}\right\} \,\middle|\, R_1 = r_1, \ldots, R_k = r_k\right) \\
&= \mathbb{P}\left(\bigcap_{j=2}^{k}\{\mu_{j,*} H_j > \theta_j \mu_{1,*} H_{1,*}\}\right) \\
&= \mathbb{E}\left[\prod_{j=2}^{k} \mathbb{P}\{H_j > \theta_{j,*} H_{1,*} \mid H_{1,*}\}\right] \\
&= \mathbb{E}\Bigg[\sum_{l_2=1}^{\infty} \cdots \sum_{l_k=1}^{\infty} \left\{\prod_{j=2}^{k} \frac{(-1)^{m_{j,l_j}} \beta_{j,l_j} c_{j,l_j}^{m_{j,l_j}+1}}{m_{j,l_j}!}\right\} \\
&\quad \times \frac{\partial^{m_{2,l_2}+\cdots+m_{k,l_k}}}{\partial(c_{2,l_2})^{m_{2,l_2}} \cdots \partial(c_{k,l_k})^{m_{k,l_k}}} \frac{\exp\left(-H_{1,*} \sum_{j=2}^{k} c_{j,l_j} \theta_{j,*}\right)}{\prod_{j=2}^{k} c_{j,l_j}}\Bigg] \\
&= \sum_{l_1=1}^{\infty} \sum_{l_2=1}^{\infty} \cdots \sum_{l_k=1}^{\infty} \left\{\prod_{j=1}^{k} \frac{(-1)^{m_{j,l_j}} \beta_{j,l_j} c_{j,l_j}^{m_{j,l_j}+1}}{m_{j,l_j}!}\right\} \\
&\quad \times \frac{\partial^{m_{2,l_2}+\cdots+m_{k,l_k}}}{\partial(c_{2,l_2})^{m_{2,l_2}} \cdots \partial(c_{k,l_k})^{m_{k,l_k}}} \frac{\int_0^{\infty} \exp\left(-x \sum_{j=1}^{k} c_{j,l_j} \theta_{j,*}\right) \mathrm{d}x}{\prod_{j=2}^{k} c_{j,l_j}} \\
&= \sum_{l_1=1}^{\infty} \sum_{l_2=1}^{\infty} \cdots \sum_{l_k=1}^{\infty} \left\{\prod_{j=1}^{k} \frac{(-1)^{m_{j,l_j}} \beta_{j,l_j} c_{j,l_j}^{m_{j,l_j}+1}}{m_{j,l_j}!}\right\} \\
&\quad \times \frac{\partial^{m_{2,l_2}+\cdots+m_{k,l_k}}}{\partial(c_{2,l_2})^{m_{2,l_2}} \cdots \partial(c_{k,l_k})^{m_{k,l_k}}} \frac{1}{\left(\sum_{n=1}^{k} c_{n,l_n} \theta_{n,*}\right) \prod_{j=2}^{k} c_{j,l_j}},
\end{aligned}
$$

from which the unconditional probability is given by

$$
\begin{aligned}
&\mathbb{P}\left(\bigcap_{j=2}^{k}\left\{\frac{P_j H_j}{R_j^{\alpha_j}} > \theta_j \frac{P_1 H_{1,*}}{R_1^{\alpha_1}}\right\}\right) \\
&= \int_{\mathbb{R}_+^k} \mathrm{d}\boldsymbol{r} f_{\boldsymbol{R}}(\boldsymbol{r})\, \mathbb{P}\left(\bigcap_{j=2}^{k}\left\{\frac{P_j H_j}{R_j^{\alpha_j}} > \theta_j \frac{P_1 H_{1,*}}{R_1^{\alpha_1}}\right\} \,\middle|\, R_1 = r_1, \ldots, R_k = r_k\right). \quad (5.82)
\end{aligned}
$$

The PMF of I is computed from (5.81), where each term in the multiple-summation is evaluated using (5.82).

Joint distribution of SINR from k "strongest" BSs when candidate serving BS in each tier is the nearest to the user

As an example of the generality of the canonical probability formulation of the joint CCDF of the SINR, we provide an outline of the derivation of the joint CCDF of the SINRs from the strongest, second-strongest, ..., kth-strongest (after selection bias) candidate serving BSs ($1 < k \le n_{\text{open}}$). Since the "strongest" (after selection bias) candidate serving BS is the serving BS for the user, this joint CCDF reduces to the marginal CCDF of the SINR (5.70) when receiving from the serving BS if $k = 1$. Suppose the serving BS selection criteria ranks the open tiers $i = 1, \ldots, n_{\text{open}}$ in decreasing order of $\tau_i Y_{B_i}$, $i = 1, \ldots, n_{\text{open}}$, as $I_1, I_2, \ldots, I_{n_{\text{open}}}$, where $I_1 \equiv I$ is the serving tier. Then the desired joint CCDF is given by

$$\mathbb{P}\{\Gamma_{I_1} > \gamma_1, \ldots, \Gamma_{I_k} > \gamma_k\} = \sum_{\substack{j_1,\ldots,j_k: \\ 1 \le j_1 < \cdots < j_k \le n_{\text{open}}}} \mathbb{P}\{\Gamma_{j_l} > \gamma_l,\ I_l = j_l,\ l = 1, \ldots, k\}$$

$$= \sum_{\substack{j_1,\ldots,j_k: \\ 1 \le j_1 < \cdots < j_k \le n_{\text{open}}}} \mathbb{P}\Big\{\Gamma_{j_l} > \gamma_l,\ l = 1, \ldots, k,\ U_{j_l} \ge \theta_{j_l}^{(j_{l+1})} U_{j_{l+1}},\ l = 1, \ldots, k-1,$$

$$U_j \le \theta_j^{(k)} U_k,\ j \in \{1, \ldots, n_{\text{open}}\} \setminus \{j_1, \ldots, j_k\}\Big\}, \quad (5.83)$$

where we have written $U_i \equiv Y_{B_i}$, $i = 1, \ldots, n_{\text{open}}$. For simplicity, we assume that the link to the user from the candidate serving BS in tier i has a fade attenuation with the same distribution as that from any other BS in tier i, $i = 1, \ldots, n_{\text{open}}$. Let us consider a typical term in (5.83). The $2k - 1$ relations in the k variables $U_{j_1}, \ldots, U_{j_k}$ given by

$$\Gamma_{j_1} > \gamma_1, \ldots, \Gamma_{j_k} > \gamma_k, U_{j_1} \ge \theta_{j_1}^{(j_2)} U_{j_2}, \ldots, U_{j_{k-1}} \ge \theta_{j_{k-1}}^{(j_k)} U_{j_k} \quad (5.84)$$

are in fact equivalent (see Problem 5.6) to the k relations in the k variables $U_{j_1}, \ldots, U_{j_k}$ given by

$$\Gamma_{j_1} > \gamma_1', \ldots, \Gamma_{j_k} > \gamma_k', \quad \gamma_l' = \frac{\rho_l'}{1 - \rho_l'},$$

$$\rho_l' = \max\{\rho_l, \theta_{j_l}^{(j_{l+1})} \rho_{l+1}\}, l = 1, \ldots, k-1,\ \gamma_k' \equiv \gamma_k. \quad (5.85)$$

With this result, we may now proceed in the same way as in the derivations of (5.73) and (5.75) to write

$$\mathbb{P}\Big\{\Gamma_{j_l} > \gamma_{j_l},\ l = 1, \ldots, k,\ U_{j_l} \ge \theta_{j_l}^{(j_{l+1})} U_{j_{l+1}},\ l = 1, \ldots, k-1,$$

$$U_j \le \theta_j^{(j_k)} U_{j_k},\ j \in \{1, \ldots, n_{\text{open}}\} \setminus \{j_1, \ldots, j_k\}\Big\}$$

$$= \mathbb{P}\Big\{\Gamma_{j_1} > \gamma_1', \ldots, \Gamma_{j_k} > \gamma_k', U_j \le \theta_j^{(j_k)} U_{j_k},\ j \in \{1, \ldots, n_{\text{open}}\} \setminus \{j_1, \ldots, j_k\}\Big\}$$

$$= \mathbb{P}\Big\{\Gamma_{j_1} > \gamma_1', \ldots, \Gamma_{j_k} > \gamma_k'\Big\} - \sum_{m=1}^{n_{\text{open}}-k} (-1)^{m-1} \sum_{\substack{j_{k+1},\ldots,j_{k+m}: \\ 1 \le j_{k+1} < \cdots < j_{k+m} \le n_{\text{open}} \\ j_{k+l} \notin \{j_1,\ldots,j_k\}, l=1,\ldots,m}}$$

$$\mathbb{P}\left(\{\Gamma_{j_1} > \gamma_1', \ldots, \Gamma_{j_k} > \gamma_k'\} \cap \bigcap_{l=1}^{m} \left\{U_{j_{k+l}} > \theta_{j_{k+l}}^{(j_k)} U_{j_k}\right\}\right). \tag{5.86}$$

Consider a typical term on the right side of (5.86). Now let $\boldsymbol{\rho}' = [\rho_1', \ldots, \rho_k']^\top$ and $\boldsymbol{\theta} = [\theta_{j_{k+1}}^{(j_k)}, \ldots, \theta_{j_{k+m}}^{(j_k)}]^\top$. Then we can show (see Problem 5.7) that

$$\mathbb{P}\left(\{\Gamma_{j_1} > \gamma_1', \ldots, \Gamma_{j_k} > \gamma_k'\} \cap \bigcap_{l=1}^{m} \left\{U_{j_{k+l}} > \theta_{j_{k+l}}^{(j_k)} U_{j_k}\right\}\right) = \mathbb{P}\{\mathbf{A}_{k+m}(\boldsymbol{\theta})\boldsymbol{X} > W\boldsymbol{b}\}, \tag{5.87}$$

where $\boldsymbol{X} = [U_{j_1}, \ldots, U_{j_{k+m}}]^\top$ and $\mathbf{A}_{k+m}(\boldsymbol{\theta})$ and $\boldsymbol{b}$ are written in block form as follows:

$$\mathbf{A}_{k+m}(\boldsymbol{\theta}) = \begin{bmatrix} \mathbf{I}_k - \boldsymbol{\rho}'\mathbf{1}_k^\top & -\boldsymbol{\rho}'\mathbf{1}_m^\top \\ -\boldsymbol{\theta}\left(\boldsymbol{e}_k^{(k)}\right)^\top & \mathbf{I}_m \end{bmatrix} \equiv \begin{bmatrix} \mathbf{A}_{11} & -\mathbf{A}_{12} \\ -\mathbf{A}_{21} & \mathbf{I}_m \end{bmatrix}, \quad \boldsymbol{b} = \begin{bmatrix} \boldsymbol{\rho}' \\ \mathbf{0}_m \end{bmatrix}. \tag{5.88}$$

Note that $\mathbf{A}_{k+m}(\boldsymbol{\theta})$ is a $(k+m) \times (k+m)$ Z-matrix. We can verify the property (2.25) for $\mathbf{A}_{k+m}(\boldsymbol{\theta})$ and use known results on the determinants and inverses of block matrices to write

$$\det \mathbf{A}_{k+m}(\boldsymbol{\theta}) = \det(\mathbf{A}_{11} - \mathbf{A}_{12}\mathbf{A}_{21}) = 1 - (\mathbf{1}_k^\top \boldsymbol{\rho}' + \rho_k' \mathbf{1}_m^\top \boldsymbol{\theta}),$$

$$\mathbf{A}_{k+m}(\boldsymbol{\theta})^{-1} = \begin{bmatrix} \mathbf{B} & \mathbf{A}_{11}^{-1}\mathbf{A}_{12}\mathbf{C} \\ \mathbf{A}_{21}\mathbf{B} & \mathbf{C} \end{bmatrix}, \quad \mathbf{A}_{11}^{-1} = \mathbf{I}_k + \frac{\boldsymbol{\rho}'\mathbf{1}_k^\top}{1 - \mathbf{1}_k^\top \boldsymbol{\rho}'},$$

$$\mathbf{B} = (\mathbf{A}_{11} - \mathbf{A}_{12}\mathbf{A}_{21})^{-1} = \mathbf{I}_k + \frac{\boldsymbol{\rho}'(\mathbf{1}_k + \mathbf{1}_m^\top \boldsymbol{\theta} \boldsymbol{e}_k^{(k)})^\top}{1 - (\mathbf{1}_k^\top \boldsymbol{\rho}' + \rho_k' \mathbf{1}_m^\top \boldsymbol{\theta})},$$

$$\mathbf{C} = (\mathbf{I}_m - \mathbf{A}_{21}\mathbf{A}_{11}^{-1}\mathbf{A}_{12})^{-1} = \mathbf{I}_m + \frac{\rho_k' \boldsymbol{\theta} \mathbf{1}_m^\top}{1 - (\mathbf{1}_k^\top \boldsymbol{\rho}' + \rho_k' \mathbf{1}_m^\top \boldsymbol{\theta})}.$$

We may then follow the same steps as in the derivation of (5.79) to arive at the desired joint CCDF.

Candidate serving BS in each tier is the "strongest" at the user

We assume, as usual, independent fading on all links from all BSs in all tiers to the user, and identically distributed fade attenuations (with PDF $f_{H_i}(\cdot)$, say) on all links from BSs in a given tier i to the user, $i = 1, \ldots, n_{\text{tier}}$. Recall that we may, without loss of generality, consider an equivalent multi-tier BS deployment modeled by the superposition PPP $\Phi = \cup_{i=1}^{n_{\text{tier}}} \Phi_i$, where all links have i.i.d. Rayleigh fading (i.e. all fade attenuations are i.i.d. Exp(1)), and where the densities $\lambda_1, \ldots, \lambda_{n_{\text{tier}}}$ of the PPPs $\Phi_1, \ldots, \Phi_{n_{\text{tier}}}$ representing the locations of the BSs in the tiers are scaled from the original densities as in (4.53).

We begin with the calculation of the terms in (5.70). Since there is exactly one candidate serving BS per tier, we can write, for each $i = 1, \ldots, n_{\text{open}}$,

$$
\begin{aligned}
&\mathbb{P}\{I = i, \Gamma_i > \gamma\} \\
&= \mathbb{P}\Big\{\Gamma_i > \gamma,\ Y_{B_j} \le \theta_j^{(i)} Y_{B_i},\ j \in \{1, \ldots, n_{\text{open}}\} \setminus \{i\}\Big\} \\
&= \sum_{b \in \Phi_i} \mathbb{P}\Big\{\Gamma_i > \gamma,\ B_i = b,\ Y_{B_j} \le \theta_j^{(i)} Y_b,\ j \in \{1, \ldots, n_{\text{open}}\} \setminus \{i\}\Big\} \\
&= \sum_{b \in \Phi_i} \mathbb{P}\Bigg\{\frac{Y_b}{N_0 + \sum_{b'' \in \Phi \setminus \{b\}} Y_{b''}} > \gamma,\quad \Big(\forall b' \in \Phi_i \setminus \{b\}\Big)\ Y_{b'} \le Y_b, \\
&\qquad \Big(\forall j \in \{1, \ldots, n_{\text{open}}\} \setminus \{i\}\Big)\ Y_{B_j} \le \theta_j^{(i)} Y_b\Bigg\} \\
&= \sum_{b \in \Phi_i} \mathbb{P}\Bigg\{\frac{Y_b}{N_0 + \sum_{b'' \in \Phi \setminus \{b\}} Y_{b''}} > \gamma\Bigg\} \\
&\quad - \sum_{b \in \Phi_i} \mathbb{P}\left(\bigcup_{b' \in \Phi_i \setminus \{b\}} \bigcup_{\substack{j=1 \\ j \ne i}}^{n_{\text{open}}} \Bigg\{\frac{Y_b}{N_0 + \sum_{b'' \in \Phi \setminus \{b\}} Y_{b''}} > \gamma, Y_{b'} > Y_b, Y_{B_j} > \theta_j^{(i)} Y_b\Bigg\}\right)
\end{aligned}
\tag{5.89}
$$

$$
\begin{aligned}
&= \sum_{k=1}^{n_{\text{open}}} (-1)^{k-1} \sum_{\substack{i_1, \ldots, i_k: \\ i_1 \equiv i \\ 1 \le i_2 < \cdots < i_k \le n_{\text{open}} \\ i_l \ne i, l = 2, \ldots, k}} \ \sum_{n_1 \ge 1, \ldots, n_k \ge 1} \frac{(-1)^{n_1 + \cdots + n_k - k}}{n_1! n_2! \cdots n_k!} \sum_{m=1}^{n_1} \sum_{\substack{b_{1,1}, \ldots, b_{1,n_1} \in \Phi_i \\ b_{2,1}, \ldots, b_{2,n_2} \in \Phi_{i_2} \\ b_{k,1}, \ldots, b_{k,n_k} \in \Phi_{i_k}}} \\
&\mathbb{P}\Bigg\{\frac{Y_{b_{1,m}}}{N_0 + \sum_{b'' \in \Phi \setminus \{b_{1,m}\}} Y_{b''}} > \gamma,\quad \Big(\forall l_1 \in \{1, \ldots, n_1\} \setminus \{m\}\Big)\ Y_{b_{1,l_1}} > Y_{b_{1,m}}, \\
&\quad \Big(\forall j \in \{2, \ldots, k\}\Big) \Big(\forall l_j \in \{1, \ldots, n_j\}\Big)\ Y_{b_{j,l_j}} > \theta_{i_j}^{(i)} Y_{b_{1,m}}\Bigg\}
\end{aligned}
\tag{5.90}
$$

$$
= \sum_{k=1}^{n_{\text{open}}} (-1)^{k-1} \sum_{\substack{i_1, \ldots, i_k: \\ i_1 \equiv i \\ 1 \le i_2 < \cdots < i_k \le n_{\text{open}} \\ i_l \ne i, l = 2, \ldots, k}} P\Big(i, i_2, \ldots, i_k; \boldsymbol{\theta}^{(i)}\Big),
\tag{5.91}
$$

where $\boldsymbol{\theta}^{(i)} = [1, \theta_{i_2}^{(i)}, \ldots, \theta_{i_k}^{(i)}]^\top$ and

$$
\begin{aligned}
P\Big(i, i_2, \ldots, i_k; \boldsymbol{\theta}^{(i)}\Big) &= \sum_{n=k}^{\infty} \frac{(-1)^{n-k}}{n!} \sum_{\substack{n_1 \ge 1, \ldots, n_k \ge 1: \\ n_1 + \cdots + n_k = n}} \binom{n}{n_1, \ldots, n_k} \sum_{m=1}^{n_1} \sum_{\substack{b_{1,1}, \ldots, b_{1,n_1} \in \Phi_i \\ b_{2,1}, \ldots, b_{2,n_2} \in \Phi_{i_2} \\ b_{k,1}, \ldots, b_{k,n_k} \in \Phi_{i_k}}} \\
&\mathbb{P}\Bigg\{\frac{Y_{b_{1,m}}}{N_0 + \sum_{b'' \in \Phi \setminus \{b_{1,m}\}} Y_{b''}} > \gamma,\quad \Big(\forall l_1 \in \{1, \ldots, n_1\} \setminus \{m\}\Big)\ Y_{b_{1,l_1}} > \theta_i^{(i)} Y_{b_{1,m}}, \\
&\qquad \Big(\forall j \in \{2, \ldots, k\}\Big) \Big(\forall l_j \in \{1, \ldots, n_j\}\Big)\ Y_{b_{j,l_j}} > \theta_{i_j}^{(i)} Y_{b_{1,m}}\Bigg\}.
\end{aligned}
\tag{5.92}
$$

Remark 5.10 Note that in (5.90) we added a "dummy" summation over the index i_1 by forcing $i_1 \equiv i$. This trick allows us to absorb the first term in (5.89) into the overall summation in (5.90) with $k = 1$ by ensuring that the multiple-summation over $i, i_2, \dots, i_k$ in (5.90) is not empty (and therefore zero) when $k = 1$. Note also the summation over m in (5.90), which arises because any of the BSs $b_{1,1}, \dots, b_{1,n_1} \in \Phi_i$ in the summation in (5.90) could be the candidate serving BS from tier i (and the overall serving BS).

We now show how to evaluate (5.91) by writing an expression for $P(1, 2, \dots, k; \boldsymbol{\theta})$, where $\boldsymbol{\theta} = [\theta_1, \theta_2, \dots, \theta_k]^\top$ with $\theta_1 \equiv 1$. Note that the expression for $P(1, 2, \dots, k; \boldsymbol{\theta})$ can be used to evaluate (5.92) by relabeling the tiers $i, i_2, \dots, i_k$. Writing $P(\boldsymbol{\theta})$ in place of $P(1, 2, \dots, k; \boldsymbol{\theta})$ for brevity, we have

$$P(\boldsymbol{\theta}) = \sum_{n=k}^{\infty} \frac{(-1)^{n-k}}{n!} \sum_{\substack{n_1 \geq 1, \dots, n_k \geq 1: \\ n_1 + \dots + n_k = n}} \binom{n}{n_1, \dots, n_k} \sum_{m=1}^{n_1} \sum_{b_{1,1} \in \Phi_1, \dots, b_{1,n_1} \in \Phi_1} \cdots \sum_{b_{k,1} \in \Phi_k, \dots, b_{k,n_k} \in \Phi_k}$$
$$\mathbb{P}\left\{ \frac{Y_{b_{1,m}}}{N_0 + \sum_{b'' \in \Phi \setminus \{b_{1,m}\}} Y_{b''}} > \gamma, \quad \left(\forall p \in \{1, \dots, n_1\} \setminus \{m\}\right) Y_{b_{1,p}} > Y_{b_{1,m}}, \right.$$
$$\left. \left(\forall j \in \{2, \dots, k\}\right) \left(\forall l \in \{1, \dots, n_j\}\right) Y_{b_{j,l}} > \theta_j Y_{b_{1,m}} \right\},$$

which closely resembles (5.26), and is evaluated in a similar way, as follows for a given $\boldsymbol{n} = [n_1, \dots, n_k]^\top$ with $n = n_1 + \dots + n_k$. Using the notation $\mathcal{N}_1, \dots, \mathcal{N}_k$ as defined in (5.27), define, for any $m \in \mathcal{N}_1 = \{1, \dots, n_1\}$, the function $g_{\boldsymbol{n},m}(\boldsymbol{x}, \mathcal{A})$ of n variables $\boldsymbol{x} = [x_1, \dots, x_n]^\top$ and a countable subset $\mathcal{A} \subseteq \mathbb{R}_+$ as follows:

$$g_{\boldsymbol{n},m}(\boldsymbol{x}, \mathcal{A}) = 1_{(\gamma,\infty)}\left(\frac{x_m}{\sum_{l \in \mathcal{N}_1 \setminus \{m\}} x_l + \sum_{l \notin \mathcal{N}_1} x_l + \sum_{y \in \mathcal{A}} y + N_0} \right)$$
$$\times \prod_{l \in \mathcal{N}_1 \setminus \{m\}} 1_{(0,\infty)}(x_l - x_m) \prod_{j=2}^{k} \prod_{l \in \mathcal{N}_j} 1_{(0,\infty)}(x_l - \theta_j x_m)$$
$$= 1_{\mathcal{U}_{\boldsymbol{n},m}\left(\sum_{y \in \mathcal{A}} y + N_0\right)}(\boldsymbol{x}),$$

where we have followed the same steps as in the derivation of (5.28). Here,

$$\mathcal{U}_{\boldsymbol{n},m}(w) = \{\boldsymbol{x} \in \mathbb{R}^n : \mathbf{A}_{\boldsymbol{n},m}(\boldsymbol{\theta}_{\boldsymbol{n}}) > w\, \boldsymbol{e}_m^{(n)}\}, \quad w > 0,$$

where, following the same steps as in the derivation of (5.76), we have

$$\mathbf{A}_{\boldsymbol{n},m}(\boldsymbol{\theta}_{\boldsymbol{n}}) = \mathbf{I}_n + [\rho^{-1} \boldsymbol{e}_m^{(n)} - \boldsymbol{\theta}_{\boldsymbol{n}}]\left(\boldsymbol{e}_m^{(n)}\right)^\top - \boldsymbol{e}_m^{(n)} \mathbf{1}_n^\top,$$
$$\rho \equiv \gamma/(1+\gamma), \quad \boldsymbol{\theta}_{\boldsymbol{n}} \equiv \left[\theta_1 \mathbf{1}_{n_1}^\top, \dots, \theta_k \mathbf{1}_{n_k}^\top\right]^\top.$$

Defining $\tilde{\Phi}_i$ to be the one-dimensional PPP of received powers at the user from the BSs in tier i, $i = 1, \ldots, n_{\text{tier}}$, and $\tilde{\Phi} = \cup_{i=1}^{n_{\text{tier}}} \tilde{\Phi}_i$, we can write

$$\sum_{b_{1,1}\in\Phi_1,\ldots,b_{1,n_1}\in\Phi_1} \cdots \sum_{b_{k,1}\in\Phi_k,\ldots,b_{k,n_k}\in\Phi_k}$$

$$\mathbb{P}\Bigg\{ \frac{Y_{b_{1,m}}}{N_0 + \sum_{b''\in\Phi\setminus\{b_{1,m}\}} Y_{b''}} > \gamma, \quad \Big(\forall p \in \{1,\ldots,n_1\} \setminus \{m\}\Big)\, Y_{b_{1,p}} > Y_{b_{1,m}},$$

$$\Big(\forall j \in \{2,\ldots,k\}\Big)\Big(\forall l \in \{1,\ldots,n_j\}\Big)\, Y_{b_{j,l}} > \theta_j Y_{b_{1,m}} \Bigg\}$$

$$= \mathbb{E}\Bigg[\sum_{\substack{y_1,\ldots,y_n\in\tilde{\Phi}: \\ y_p \neq y_q \text{ for } p\neq q \\ y_l\in\tilde{\Phi}_j,\, l\in\mathcal{N}_j,\, j=1,\ldots,k}} g_{\boldsymbol{n},m}(\boldsymbol{y}, \tilde{\Phi} \setminus \{y_1,\ldots,y_n\}) \Bigg]$$

$$= \int\cdots\int_{\mathbb{R}_{++}^n} \left[\prod_{i=1}^{k} \left(\frac{2\pi\lambda_i}{P_i}\right)^{n_i} \prod_{l\in\mathcal{N}_i} r_l^{1+\alpha_i} \right]$$

$$\mathbb{E}\left[\int_{\mathcal{U}_{\boldsymbol{n},m}(W)\cap\mathbb{R}_{++}^n} \prod_{i=1}^{k} \prod_{l\in\mathcal{N}_i} \exp\left(-\frac{r_l^{\alpha_i}}{P_i} x_l\right) \mathrm{d}\boldsymbol{x} \right] \mathrm{d}r_1 \cdots \mathrm{d}r_n,$$

where we have followed the same steps as in the derivation of (5.33), and, as usual, W is the thermal noise plus the total received power at the user from all BSs in all tiers.

Then the same steps as in the derivation of (5.35) yield the result

$$\left(\prod_{i=1}^{k} \prod_{l\in\mathcal{N}_i} \frac{r_l^{\alpha_i}}{P_i} \right) \mathbb{E}\left[\int_{\mathcal{U}_{\boldsymbol{n},m}(W)\cap\mathbb{R}_{++}^n} \prod_{i=1}^{k} \prod_{l\in\mathcal{N}_i} \exp\left(-\frac{r_l^{\alpha_i}}{P_i} x_l\right) \mathrm{d}\boldsymbol{x} \right]$$

$$= \begin{cases} 0, & \gamma \geq (\boldsymbol{n}^\top \boldsymbol{\theta} - 1)^{-1}, \\ \det \mathbf{A}_{\boldsymbol{n},m}(\boldsymbol{\theta}_{\boldsymbol{n}})^{-1} \displaystyle\prod_{i=1}^{k} \prod_{l\in\mathcal{N}_i} \frac{r_l^{\alpha_i}/P_i}{\left[\frac{r_1^{\alpha_{\boldsymbol{n},1}}}{P_{\boldsymbol{n},1}}, \ldots, \frac{r_n^{\alpha_{\boldsymbol{n},n}}}{P_{\boldsymbol{n},n}}\right]^\top \mathbf{A}_{\boldsymbol{n},m}(\boldsymbol{\theta}_{\boldsymbol{n}})^{-1} \boldsymbol{e}_l^{(n)}} & \\ \times \mathcal{L}_W\left(\left[\frac{r_1^{\alpha_{\boldsymbol{n},1}}}{P_{\boldsymbol{n},1}}, \ldots, \frac{r_n^{\alpha_{\boldsymbol{n},n}}}{P_{\boldsymbol{n},n}}\right]^\top \mathbf{A}_{\boldsymbol{n},m}(\boldsymbol{\theta}_{\boldsymbol{n}})^{-1} \boldsymbol{e}_m^{(n)} \right), & \gamma < (\boldsymbol{n}^\top \boldsymbol{\theta} - 1)^{-1}, \end{cases}$$

where $\mathcal{L}_W(\cdot)$ is given by (5.36), and, following (5.77) and (5.78), we can write

$$\det \mathbf{A}_{\boldsymbol{n},m}(\boldsymbol{\theta}_{\boldsymbol{n}}) = \mathbf{1}_n^\top \mathbf{A}_{\boldsymbol{n},m}(\boldsymbol{\theta}_{\boldsymbol{n}}) \boldsymbol{e}_m^{(n)} = \rho^{-1} - \mathbf{1}_n^\top \boldsymbol{\theta}_{\boldsymbol{n}} = \rho^{-1} - \boldsymbol{n}^\top \boldsymbol{\theta},$$

$$\mathbf{A}_{\boldsymbol{n},m}(\boldsymbol{\theta}_{\boldsymbol{n}})^{-1} = \mathbf{I}_n + \frac{\boldsymbol{\theta}_{\boldsymbol{n}} \mathbf{1}_n^\top}{\rho^{-1} - \boldsymbol{n}^\top \boldsymbol{\theta}} - \boldsymbol{e}_m^{(n)} \left(\boldsymbol{e}_m^{(n)}\right)^\top,$$

$$\mathbf{A}_{\boldsymbol{n},m}(\boldsymbol{\theta}_{\boldsymbol{n}})^{-1} \boldsymbol{e}_l^{(n)} = (1 - \delta_{m,l}) \boldsymbol{e}_l^{(n)} + \frac{\boldsymbol{\theta}_{\boldsymbol{n}}}{\rho^{-1} - \boldsymbol{n}^\top \boldsymbol{\theta}}, \quad l = 1, \ldots, n.$$

Finally, if we make the change of variables

$$u_l = \left(\frac{\theta_i}{\rho^{-1} - \boldsymbol{n}^\top \boldsymbol{\theta}}\right)^{2/\alpha_i} \pi\lambda_i r_l^2, \quad l \in \mathcal{N}_i, \quad i = 1, \ldots, k,$$

we can write

$$P(\boldsymbol{\theta}) = \sum_{n=k}^{\infty} \frac{(-1)^{n-k}}{n!} \sum_{\substack{n_1 \geq 1, \ldots, n_k \geq 1: \\ n_1 + \cdots + n_k = n \\ n_1\theta_1 + \cdots + n_k\theta_k < \rho^{-1}}} \binom{n}{n_1, \ldots, n_k}$$

$$\times \frac{1}{\rho^{-1} - \boldsymbol{n}^\top \boldsymbol{\theta}} \left[\prod_{i=1}^{k} \left(\frac{\rho^{-1} - \boldsymbol{n}^\top \boldsymbol{\theta}}{\theta_i}\right)^{2n_i/\alpha_i}\right] \sum_{m=1}^{n_1}$$

$$\int_{\mathbb{R}_{++}^n} \frac{\exp\left[-N_0 \sum_{i=1}^{k} \frac{\sum_{l \in \mathcal{N}_i} u_l^{\alpha_i/2}}{(\pi\lambda_i)^{\alpha_i/2} P_i} - \pi \sum_{j=1}^{n_{\text{tier}}} \frac{\lambda_j P_j^{2/\alpha_j}}{\operatorname{sinc}(2/\alpha_j)} \left(\sum_{i=1}^{k} \frac{\sum_{l \in \mathcal{N}_i} u_l^{\alpha_i/2}}{(\pi\lambda_i)^{\alpha_i/2} P_i}\right)^{2/\alpha_j}\right]}{\prod_{i=1}^{k} \prod_{l \in \mathcal{N}_i} \left[(1 - \delta_{m,l}) + \frac{\theta_i (\pi\lambda_i)^{\alpha_i/2} P_i}{(\rho^{-1} - \boldsymbol{n}^\top \boldsymbol{\theta}) u_l^{\alpha_i/2}} \sum_{j=1}^{k} \frac{\sum_{p \in \mathcal{N}_j} u_p^{\alpha_j/2}}{(\pi\lambda_j)^{\alpha_j/2} P_j}\right]} \, d\boldsymbol{u}. \tag{5.93}$$

Finally we obtain the following theorem.

THEOREM 5.12 *(Mukherjee, 2012b.) Suppose we are given n_{tier} tiers of BSs located at the points of independent homogeneous PPPs with densities $\nu_1, \ldots, \nu_{n_{\text{tier}}}$. Suppose the candidate serving BSs from the open tiers are chosen as the BSs received most strongly at the user, and the fade attenuation on the links from all the BSs in tier i are i.i.d. with PDF $f_{H_i}(\cdot)$, $i = 1, \ldots, n_{\text{tier}}$. If the serving BS is the candidate serving BS from the serving tier $I = \arg\max_{1 \leq i \leq n_{\text{open}}} \tau_i Y_{B_i}$, then the CCDF of the SINR at the user when receiving from this serving BS, conditioned on the distances to the candidate serving BSs in all open tiers, is given by* (5.70)*. Each term on the right side of* (5.70) *is calculated from* (5.91)*, each term of which in turn is calculated using* (5.93)*, where $\lambda_i = \nu_i \, \mathbb{E}[H_i^{2/\alpha_i}] / \Gamma(1 + 2/\alpha_i)$, $i = 1, \ldots, n_{\text{tier}}$.*

Remark 5.11 It is an interesting exercise to prove that, for a single tier ($k = 1 = n_{\text{open}} = n_{\text{tier}}$, $\theta_1 \equiv 1$), Theorem 5.12 yields the same result (4.72) for the distribution of the maximum SINR at the user from the BSs in the tier as previously obtained by direct calculation in Theorem 4.16 (see Problem 5.8).

The distribution of the serving tier proves to be surprisingly easy to obtain, using the analysis in Section 5.3.1, as follows. Consider the one-dimensional PPPs of received powers from the BSs in the tiers of a multi-tier deployment with BSs located at the points of independent homogeneous PPPs with densities $\nu_1, \ldots, \nu_{n_{\text{tier}}}$ for the tiers, and arbitrary i.i.d. fading with PDF $f_{H_i}(\cdot)$ on all links to the user from BSs in tier i, $i = 1, \ldots, n_{\text{tier}}$. Recall from Theorem 4.1 that these one-dimensional PPPs are, respectively, equivalent to the one-dimensional PPPs of received powers from BSs of a different multi-tier deployment, where all BSs in all tiers transmit with the same powers as before, the path-loss model on all links is the same, but there is no fading, and the tier densities are given by $\lambda_i = \nu_i \mathbb{E}[H_i^{2/\alpha_i}]$, $i = 1, \ldots, n_{\text{tier}}$, where H_i is a random variable with PDF $f_{H_i}(\cdot)$ and α_i is the path-loss exponent on the links to the user from BSs in tier i in both deployments, $i = 1, \ldots, n_{\text{tier}}$. In the new deployment with no fading, the "strongest" BS in each tier is simply the one that is "nearest" to the user, and the received power Y_{B_i} from the "strongest" BS in tier i, say, is then $P_i/R_i^{\alpha_i}$, where R_i is the distance of the nearest BS in tier i from the user. Then the serving tier selection criterion $I = \arg\max_{1\leq i\leq n_{\text{open}}} \tau_i Y_{B_i}$ becomes

$$I = \arg \max_{1\leq i\leq n_{\text{open}}} \frac{\tau_i P_i}{R_i^{\alpha_i}} = \arg \min_{1\leq i\leq n_{\text{open}}} \tilde{\tau}_i R_i^{\alpha_i}, \quad \tilde{\tau}_i = \frac{1}{\tau_i P_i}, \quad i = 1, \ldots, n_{\text{open}}.$$

In other words, the PMF of I is given by (5.67), with $\theta_j^{(i)} = \tau_i/\tau_j$ replaced by $\tilde{\theta}_j^{(i)} = \tilde{\tau}_i/\tilde{\tau}_j = (\tau_j/\tau_i)(P_j/P_i), j = 1, \ldots, n_{\text{open}}$:

$$\mathbb{P}\{I = i\} = \int_0^\infty \exp\left\{-\pi \sum_{j=1}^{n_{\text{open}}} \lambda_j (\tau_j P_j)^{2/\alpha_j} \left[\frac{u_i}{\pi \lambda_i (\tau_i P_i)^{2/\alpha_i}}\right]^{\alpha_i/\alpha_j}\right\} \mathrm{d}u_i,$$
$$i = 1, \ldots, n_{\text{open}}. \tag{5.94}$$

This result was first derived in Madhusudhanan *et al.* (2012a).

Remark 5.12 Since the joint CCDF of the SINRs at the user when receiving from the candidate serving BSs in the tiers is a function of only N_0 and $\lambda_i^{\alpha_i/2} K_i P_i^{\text{tx}}$, $i = 1, \ldots, n_{\text{tier}}$, it is not surprising that this is also true of the SINR from the serving BS. This can also be directly verified from (5.93) and (5.94).

Problems

5.4 Prove (5.76). *Hint*: Follow the hint in Problem 2.4 for the first inequality.

5.5 Prove that $\mathbf{A}_k(\boldsymbol{\theta})$ satisfies (2.25) and calculate det $\mathbf{A}_k(\boldsymbol{\theta})$. *Hint*: Calculate the determinant of any principal submatrix of $\mathbf{A}_k(\boldsymbol{\theta})$ by replacing the first row of this submatrix by the sum of all the rows of the submatrix (which does not change the determinant), then expanding the determinant along the first row.

5.6 Prove that the conditions in (5.84) and (5.85) are equivalent. *Hint*: Let $W(j_1, \ldots, j_k)$ be the total interference at the user from all BSs in all tiers other than $\{j_1, \ldots, j_k\}$, and define $\boldsymbol{U} = [U_{j_1}, \ldots, U_{j_k}]^\top$. Then, for any $l = 1, \ldots, k$, $\Gamma_{j_l} > \gamma_l \Leftrightarrow U_{j_l} > \rho_l[\mathbf{1}_k^\top \boldsymbol{U} + W(j_1, \ldots, j_k)]$, and so, for $l = 1, \ldots, k-1$, the two conditions $\Gamma_{j_l} > \gamma_l$ and $U_{j_l} \geq \theta_{j_l}^{(j_{l+1})} U_{j_{l+1}}$ are equivalent to the single condition

$$U_{j_l} > \max\left\{\rho_l, \rho_{l+1}\theta_{j_l}^{(j_{l+1})}\right\}[\mathbf{1}_k^\top \boldsymbol{U} + W(j_1, \ldots, j_k)],$$

i.e. $\Gamma_{j_l} > \gamma_l'$, where γ_l', $l = 1, \ldots, k-1$, are as defined in (5.85).

5.7 Verify (5.87), where $\mathbf{A}_{k+m}(\boldsymbol{\theta})$ and $\boldsymbol{b}$ are as defined in (5.88).

5.8 Prove the statement in Remark 5.11. *Hint*: For a single-tier PPP Φ, from (4.55) and (4.56) we have

$$\mathbb{P}\{\Gamma > \gamma\} = \sum_{n=1}^{\infty} \frac{(-1)^{n-1}}{n!} \sum_{\substack{b_1,\ldots,b_n \in \Phi: \\ (\forall i \neq j)\, b_i \neq b_j}} \mathbb{P}\left\{ \frac{Y_{b_l}}{N_0 + \sum_{b \in \Phi \setminus \{b_l\}} Y_b} > \gamma,\ l = 1,\ldots,n \right\}$$

$$= \sum_{n=1}^{\infty} \frac{(-1)^{n-1}}{n!} \sum_{\substack{b_1,\ldots,b_n \in \Phi: \\ (\forall i \neq j)\, b_i \neq b_j}} \sum_{m=1}^{n}$$

$$\mathbb{P}\left\{ Y_{b_m} = \min_{1 \leq l \leq n} Y_{b_l},\ \frac{Y_{b_l}}{N_0 + \sum_{b \in \Phi \setminus \{b_l\}} Y_b} > \gamma,\ l = 1,\ldots,n \right\}$$

$$= \sum_{n=1}^{\infty} \frac{(-1)^{n-1}}{n!} \sum_{m=1}^{n} \sum_{\substack{b_1,\ldots,b_n \in \Phi: \\ (\forall i \neq j)\, b_i \neq b_j}}$$

$$\mathbb{P}\left\{ \frac{Y_{b_m}}{N_0 + \sum_{b \in \Phi \setminus \{b_m\}} Y_b} > \gamma,\ (\forall l \in \{1,\ldots,n\} \setminus \{m\})\, Y_{b_l} > Y_{b_m} \right\},$$

which is just (5.90) for a single tier ($k \equiv 1$, $n_1 \equiv n$, $\Phi_i \equiv \Phi$).

5.3.3 Serving BS is the max-SINR (after selection bias) candidate serving BS

In cellular networks today, the selection of the BS to serve a user is made on the basis of received power, not the SINR, so this serving BS selection criterion is not a good model for any of today's deployed cellular networks. Observe that if the candidate serving BS from each tier is the one that is received at the user with maximum power (i.e. maximum SINR) among all BSs in that tier, then, in the absence of any selection bias, the max-SINR serving BS selection criterion yields the BS that is received with maximum SINR at the user among all BSs in the network. However, if selection bias is applied to the SINRs when receiving from the candidate serving BSs, the selected serving BS may not necessarily yield the maximum SINR at the user among all BSs in the network. Nonetheless, we provide a brief discussion of this selection criterion here for the sake of completeness.

The serving BS is the candidate serving BS of tier $I = \arg\max_{1 \leq i \leq n_{\text{open}}} \tau_i \Gamma_i$. As before, we may write

$$\mathbb{P}\{\Gamma_I > \gamma\} = \sum_{i=1}^{n_{\text{open}}} \mathbb{P}\{\Gamma_i > \gamma, I = i\} = \sum_{i=1}^{n_{\text{open}}} \mathbb{P}\{\Gamma_i > \gamma,$$
$$(\forall j \in \{1,\ldots,n_{\text{open}}\} \setminus \{i\})\ \tau_j \Gamma_j \leq \tau_i \Gamma_i\}.$$

Unfortunately, a relation of the form $\tau_j \Gamma_j \leq \tau_i \Gamma_i$ is nonlinear in the received powers, so our earlier approach of framing the CCDF of the SINR as a canonical probability problem involving the vector of received powers from the candidate serving BSs in the open tiers will not work here. Instead, let us work with the random variables $\Gamma_1, \ldots, \Gamma_{n_{\text{open}}}$ directly as follows. First, for each $i = 1, \ldots, n_{\text{open}}$, we can write

$$
\begin{aligned}
&\mathbb{P}\{\Gamma_i > \gamma,\ (\forall j \in \{1,\dots,n_{\text{open}}\} \setminus \{i\})\ \tau_j \Gamma_j \le \tau_i \Gamma_i\} \\
&= \mathbb{P}\{\Gamma_i > \gamma\} - \sum_{k=2}^{n_{\text{open}}} (-1)^k \sum_{\substack{i_2,\dots,i_k: \\ 1 \le i_2 < \dots < i_k: \\ i_l \ne i, l=2,\dots,k}} \mathbb{P}\left(\{\Gamma_i > \gamma\} \cap \bigcup_{l=2}^{k} \{\Gamma_{i_l} > \theta_{i_l}^{(i)} \Gamma_i\} \right),
\end{aligned}
$$

and, for any $i_2, \dots, i_k$ in this equation, we see that

$$
\mathbb{P}\left(\{\Gamma_i > \gamma\} \cap \bigcup_{l=2}^{k} \{\Gamma_{i_l} > \theta_{i_l}^{(i)} \Gamma_i\} \right) = \mathbb{P}\{\mathbf{A}\boldsymbol{\Gamma} > \tilde{\boldsymbol{b}}\},
$$

where $\boldsymbol{\Gamma} = [\Gamma_i, \Gamma_{i_2}, \dots, \Gamma_{i_k}]^\top$ and

$$
\mathbf{A} = \mathbf{I}_k - [1, \theta_{i_2}^{(i)}, \dots, \theta_{i_k}^{(i)}]^\top \left(\boldsymbol{e}_1^{(k)}\right)^\top + \boldsymbol{e}_1^{(k)} \left(\boldsymbol{e}_1^{(k)}\right)^\top, \quad \tilde{\boldsymbol{b}} = \gamma \boldsymbol{e}_1^{(k)}.
$$

Again, we can verify that $\mathbf{A}$ is a Z-matrix satisfying (2.25). We cannot apply Lemma 2.3 directly because $\Gamma_{i_1}, \dots, \Gamma_{i_k}$ are not independent. However, Lemma 2.3 can still be applied provided we first express the *joint PDF* of $\boldsymbol{\Gamma}$ as a *tensor product* of univariate Erlang PDFs:

$$
f_{\boldsymbol{\Gamma}}(x_1, \dots, x_k) = \sum_{j_1 \ge 1, \dots, j_k \ge 1} \beta_{j_1,\dots,j_k} \prod_{l=1}^{k} g(x_l; m_{j_l,l}, c_{j_l,l}), \qquad \sum_{j_1 \ge 1, \dots, j_k \ge 1} \beta_{j_1,\dots,j_k} = 1,
$$

where each $g(x; m, c)$ is of the form (2.18). Of course, it is far from straightforward to find such a representation of the joint PDF of $\boldsymbol{\Gamma}$.

If n_{open} is small, we may instead opt to compute the CCDF directly in terms of the appropriate PDFs, as follows:

$$
\begin{aligned}
\mathbb{P}\{\Gamma > \gamma\} &= 1 - \sum_{i=1}^{n_{\text{open}}} \mathbb{P}\left\{\Gamma_i \le \gamma,\ \Gamma_j \le \theta_j^{(i)} \Gamma_i,\quad j \in \{1, \dots, n_{\text{open}}\} \setminus \{i\}\right\} \\
&= 1 - \sum_{i=1}^{n_{\text{open}}} \int_0^{\gamma} \mathrm{d}\gamma_i \int_0^{\theta_1^{(i)} \gamma_i} \mathrm{d}\gamma_1 \cdots \int_0^{\theta_{i-1}^{(i)} \gamma_i} \mathrm{d}\gamma_{i-1} \\
&\quad \times \int_0^{\theta_{i+1}^{(i)} \gamma_i} \mathrm{d}\gamma_{i+1} \cdots \int_0^{\theta_{n_{\text{open}}}^{(i)} \gamma_i} \mathrm{d}\gamma_{n_{\text{open}}} f_{\Gamma_1,\dots,\Gamma_{n_{\text{open}}}}(\gamma_1, \dots, \gamma_{n_{\text{open}}}),
\end{aligned}
$$

where the joint PDF of $\Gamma_1, \dots, \Gamma_{n_{\text{open}}}$ is given by

$$
f_{\Gamma_1,\dots,\Gamma_{n_{\text{open}}}}(\gamma_1, \dots, \gamma_{n_{\text{open}}}) = \frac{(-1)^{n_{\text{open}}} \partial^{n_{\text{open}}}}{\partial \gamma_1 \cdots \partial \gamma_{n_{\text{open}}}} \mathbb{P}\{\Gamma_1 > \gamma_1, \dots, \Gamma_{n_{\text{open}}} > \gamma_{n_{\text{open}}}\}
$$

and $\mathbb{P}\{\Gamma_1 > \gamma_1, \dots, \Gamma_{n_{\text{open}}} > \gamma_{n_{\text{open}}}\}$ is given by (5.8) and (5.12) for the "nearest BS" candidate serving BS rule, and by (5.38) for the "strongest BS" candidate serving BS selection rule. A similar direct computation is also possible for the PMF of the serving tier

$$
\mathbb{P}\{I = i\} = \mathbb{P}\{(\forall j \in \{1, \dots, n_{\text{open}}\} \setminus \{i\})\ \tau_j \Gamma_j \le \tau_i \Gamma_i\}, \quad i = 1, \dots, n_{\text{open}}.
$$

Application: macro-pico HCN

Consider an HCN with $n_{\text{tier}} = 2 = n_{\text{open}}$ tiers, tier 1 being the macro and tier 2 the pico. Neglect thermal noise and assume the same path-loss exponent $\alpha = 4$ and i.i.d. Rayleigh fading on all links. We drop the subscript 1 from macro-tier quantities and replace the subscript 2 with a prime for pico-tier quantities. Consider an arbitrarily located UE in such an HCN. For simplicity, we assume the candidate serving BSs from the macro and pico tiers are the ones closest to the UE. From our previous results, it is easy to see that the joint CCDF at the UE from the nearest macro and nearest pico BS is given by (Mukherjee, 2011b)

$$\mathbb{P}\{\Gamma > \gamma, \Gamma' > \gamma' \mid R = r, R' = r'\} = \frac{(1-\gamma\gamma')\mathcal{L}_W\left(\frac{1}{1-\gamma\gamma'}\left[\frac{\gamma(1+\gamma')r^4}{P} + \frac{\gamma'(1+\gamma)(r')^4}{P'}\right]\right)}{\left[1+\gamma\frac{P'}{P}\left(\frac{r}{r'}\right)^4\right]\left[1+\gamma'\frac{P}{P'}\left(\frac{r'}{r}\right)^4\right]}, \quad \gamma\gamma' < 1, \tag{5.95}$$

where

$$\mathcal{L}_W(s) = \exp\left\{-\pi\left[\lambda\sqrt{Ps}\cot^{-1}\left(\frac{r^2}{\sqrt{Ps}}\right) + \lambda'\sqrt{P's}\cot^{-1}\left(\frac{(r')^2}{\sqrt{P's}}\right)\right]\right\}, \quad s > 0.$$

Here, the joint PDF may be obtained by algebraic differentiation of the joint CCDF:

$$f_{\Gamma,\Gamma'|R,R'}(\gamma,\gamma'|r,r') = \frac{\partial^2}{\partial\gamma\,\partial\gamma'}\mathbb{P}\{\Gamma > \gamma, \Gamma' > \gamma' \mid R = r, R' = r'\}, \quad \gamma\gamma' < 1.$$

In fact, by writing the joint CCDF in the form

$$\mathbb{P}\{\Gamma > \gamma, \Gamma' > \gamma' \mid R = r, R' = r'\} = P_1(\gamma,\gamma';\tilde{a})P_2(s(\gamma,\gamma';\tilde{a});\tilde{a},\tilde{\mu},B), \quad \gamma\gamma' < 1,$$

where

$$\begin{aligned}
&a = r^4/P, && a' = (r')^4/P, && A = a + a', \\
&\tilde{a} = a/A = 1/[1+(P/P')(r'/r)^4], && m = \lambda\pi r^2, && m' = \lambda'\pi(r')^2, \\
&\mu = \pi\lambda\sqrt{AP} = m/\sqrt{\tilde{a}}, && \mu' = \pi\lambda'\sqrt{AP'} = m'/\sqrt{1-\tilde{a}}, && B = \mu + \mu', \\
&\tilde{\mu} = \mu/B = 1/[1+(\lambda'/\lambda)\sqrt{P'/P}],
\end{aligned}$$

and

$$P_1(\gamma,\gamma';\tilde{a}) = \frac{1-\gamma\gamma'}{\left(1+\gamma\frac{a}{a'}\right)\left(1+\gamma'\frac{a'}{a}\right)} = \frac{1}{1+\gamma\frac{\tilde{a}}{1-\tilde{a}}} + \frac{1}{1+\gamma'\frac{1-\tilde{a}}{\tilde{a}}} - 1,$$

$$P_2(s;\tilde{a},\tilde{\mu},B) = \exp\left\{B\left[g(s;\sqrt{\tilde{a}},\tilde{\mu}) + g(s;\sqrt{1-\tilde{a}},1-\tilde{\mu})\right]\right\},$$

$$s(\gamma,\gamma';\tilde{a}) = \sqrt{\frac{\gamma(1+\gamma')\tilde{a} + \gamma'(1+\gamma)(1-\tilde{a})}{1-\gamma\gamma'}}$$

$$= \sqrt{\frac{(1+\gamma)\tilde{a} + (1+\gamma')(1-\tilde{a})}{1-\gamma\gamma'}} - 1,$$

$$g(s; b, \nu) = -\nu s \cot^{-1}\frac{b}{s}, \quad s > 0,\ 0 < \nu < 1,\ 0 < b < 1,$$

we obtain, after some manipulation,

$$f_{\Gamma,\Gamma'|R,R'}(\gamma,\gamma'|r,r')$$
$$= P_2P_1\left\{h\left[\left(\frac{\partial^2 s}{\partial\gamma\,\partial\gamma'} + h\frac{\partial s}{\partial\gamma}\frac{\partial s}{\partial\gamma'}\right) + \frac{1}{P_1}\left(\frac{\partial P_1}{\partial\gamma}\frac{\partial s}{\partial\gamma'} + \frac{\partial s}{\partial\gamma}\frac{\partial P_1}{\partial\gamma'}\right)\right]\right.$$
$$\left. + \frac{\partial h}{\partial s}\frac{\partial s}{\partial\gamma}\frac{\partial s}{\partial\gamma'}\right\}, \ \gamma\gamma' < 1,$$

where

$$h(s; \tilde{a}, \tilde{\mu}) = \frac{1}{s}\ln P_2 - Bs\left[\frac{\tilde{\mu}\sqrt{\tilde{a}}}{s^2+\tilde{a}} + \frac{(1-\tilde{\mu})\sqrt{1-\tilde{a}}}{s^2+1-\tilde{a}}\right],$$

and the various partial derivatives may be evaluated directly:

$$\frac{\partial P_1}{\partial\gamma} = -\frac{\tilde{a}(1-\tilde{a})}{[\tilde{a}\gamma + (1-\tilde{a})]^2}, \qquad \frac{\partial P_1}{\partial\gamma'} = -\frac{\tilde{a}(1-\tilde{a})}{[(1-\tilde{a})\gamma' + \tilde{a}]^2},$$

$$\frac{\partial s}{\partial\gamma} = \frac{\left[s - \dfrac{\gamma'(1-\tilde{a})}{s}\right]}{2\gamma(1-\gamma\gamma')}, \qquad \frac{\partial s}{\partial\gamma'} = \frac{\left(s - \dfrac{\gamma\tilde{a}}{s}\right)}{2\gamma'(1-\gamma\gamma')},$$

$$\frac{\partial^2 s}{\partial\gamma\,\partial\gamma'} = \frac{\left[3s + \dfrac{1}{s} - \dfrac{\tilde{a}(1-\tilde{a})}{s^3}\right]}{4(1-\gamma\gamma')^2}, \qquad \frac{\partial g(s;b,\nu)}{\partial s} = \frac{g(s;b,\nu)}{s} - \nu\frac{bs}{s^2+b^2},$$

$$\frac{\partial^2 g(s;b,\nu)}{\partial s^2} = -\frac{2\nu b^3}{(s^2+b^2)^2}, \qquad \frac{\partial h}{\partial s} = -2B\left[\frac{\tilde{\mu}\tilde{a}^{3/2}}{(s^2+\tilde{a})^2} + \frac{(1-\tilde{\mu})(1-\tilde{a})^{3/2}}{(s^2+1-\tilde{a})^2}\right].$$

Note that P_1, P_2, and s are all positive, whereas g is always negative, and the parameters $\tilde{a}$ and $\tilde{\mu}$ take values between 0 and 1. This formulation provides numerical stability for computation.

If the UE is served by the stronger of the two nearest BSs (one macro, one pico) with selection bias of $\tau > 1$ in favor of the pico tier, the probability that a UE located at a distance of $R = r$ from the nearest macro BS and $R' = r'$ from the nearest pico BS is served by the macro BS is (Mukherjee & Güvenç, 2011)

$$\mathbb{P}\{\Gamma > \tau\Gamma' \mid R = r, R' = r'\} = \int_0^{1/\sqrt{\tau}}\int_{\tau\gamma'}^{1/\gamma'} f_{\Gamma,\Gamma'|R,R'}(\gamma,\gamma'|r,r')\mathrm{d}\gamma\,\mathrm{d}\gamma'.$$

5.4 Selection bias and the need for interference control

Let us consider the serving BS selection scheme given by $I = \arg\max_{1\le i\le n_{\text{open}}} \tau_i Y_{B_i}$. Note that, without loss of generality (see Problem 5.9), we may take $\min_{1\le i\le n_{\text{open}}}$

$\tau_i = 1$, so $\tau_i \geq 1$, $i = 1, \ldots, n_{\text{open}}$. Suppose also that we relabel the tiers so that $\arg\min_{1 \leq i \leq n_{\text{open}}} \tau_i = 1$. In other words, tier 1 is the baseline, and every other open tier $j > 1$ has preference (through a factor $\tau_j > 1$) when it comes to serving the user. This is the method presently employed in macro-pico HCNs, where tier 1 is the macro tier and tier 2 is the pico tier. The objective of this selection-bias-based serving BS selection scheme, called *cell range expansion* (CRE) in LTE in the context of a macro-pico HCN, is to *offload* UEs from the macro tier onto the pico tier by getting them to associate preferentially with (be served by) the pico tier. The larger the factor τ_2 (called the *range expansion bias* or REB in LTE), the greater the fraction of UEs in the system that is served by the pico BSs instead of macro BSs.

However, this offloading of UEs from the macros to the picos comes at a cost. Observe that the received power from the candidate serving pico BS was "boosted" by the factor $\tau_2 > 1$ for the purposes of serving BS selection, but that the SINR at the UE is computed from the actual (not boosted) received powers from the candidate serving pico BS and the candidate serving macro BS, and is not a function of the REB τ_2. It follows that, if τ_2 is chosen high enough, then, with high probability, the received power Y_{B_1} from the candidate serving macro BS does not exceed the received power Y_{B_2} from the candidate serving pico BS after it is "boosted" by the factor τ_2, i.e. the event $\{Y_{B_1} \leq \tau_2 Y_{B_2}\}$ occurs with high probability. Then, with high probability, the serving tier is $I = 2$, i.e. the pico tier. However, $\{Y_{B_1} > Y_{B_2}\}$, the event that the received power from the macro candidate serving BS exceeds that from the pico candidate serving BS, also occurs with high probability given the typical difference in transmit powers between the macro and pico BSs. Then $\Gamma_1 > \Gamma_2$, i.e. the UE is served by the tier from which it receives lower than optimal SINR. In fact, the SINR at the UE when receiving from the pico serving BS may be poor because it suffers strong interference from the candidate serving macro BS that it rejected in order to associate with the pico tier. Calculations of, say, $\mathbb{P}\{\Gamma_1 > \Gamma_2 \mid R_1 = r_1, R_2 = r_2\}$, where R_1 and R_2 are the distances to the candidate serving BSs in the macro and pico tiers, respectively, can show that this problem of poor SINR at a user served by the pico tier selected after high REB can become a serious issue in a macro-pico HCN.

The preceding discussion is a consequence of a fundamental aspect of HCNs (as opposed to other heterogeneous networks such as wireless sensor networks, for example): they are almost always *interference limited*, meaning that the CCDF of the SINR (when receiving from the serving BS) is limited not by thermal noise but by the interference from non-serving BSs in all tiers. Thus, the distribution of the SINR at the user can be improved by *interference control*. Measures to reduce interference without communication between the BSs are usually termed *interference mitigation*, while methods that employ communication amongst the BSs of the tiers are called *interference coordination*.

We study one kind of interference control method in this book, namely *power control*, which directly attacks interference by modifying the transmit powers of the interferers. For example, in the macro-pico HCN example, since the problem of poor SINR at UEs served by the pico tier is caused by the high transmit power of the macro BSs relative to the pico BSs, it may be alleviated by reducing the transmit power of the macro BSs.

Of course, since the macro tier was designed to satisfy certain coverage requirements, it is not possible to reduce the transmit power of the macro BSs permanently. Instead, the solution proposed in LTE is to specify certain time intervals during which the macro BSs transmit with as little power as possible (ideally, no power at all on the traffic channels – of course, the pilot and control channels must continue to be powered as before, even during these intervals). Thus, during such an interval (called an *almost-blank subframe* or ABS in LTE), the UEs served by pico BSs experience a respite from macro-tier interference and should be able to enjoy relatively high SINR, permitting high data rates. In LTE, this solution, using ABSs for macro BSs, is called "enhanced inter-cell interference coordination" or eICIC. However, this is only one particular form of power control. We explore this and other power control schemes, and their impact on the distribution of the SINR at an arbitrary user in the network, in Chapter 6.

Problem

5.9 Prove that in the serving tier selection rule $I = \arg\max_{1 \le i \le n_{\text{open}}} \tau_i Y_{B_i}$, we may assume $\tau_i \ge 1$, $i = 1, \ldots, n_{\text{open}}$, without loss of generality. *Hint*: As $\tau_i > 0$, $i = 1, \ldots, n_{\text{open}}$, we can write

$$I = \arg \max_{1 \le i \le n_{\text{open}}} \tau_i Y_{B_i} = \arg \max_{1 \le i \le n_{\text{open}}} \frac{\tau_i}{\min_{1 \le j \le n_{\text{open}}} \tau_j} Y_{B_i}.$$

6 SINR analysis with power control

6.1 Introduction

In Chapters 4 and 5, we analyzed the distribution of the SINR at a user in single-tier and multi-tier deployments, respectively, when all BSs in each tier transmitted with the same, fixed, power. This is a good model for the pilot channels in a system such as LTE, because the pilot channels are used by the UEs to identify BSs, and it makes sense for the power in these pilot channels to be as high as possible all the time. We also know that the distribution of maximum SIR (without selection bias) in an interference-limited deployment is not a function of the (assumed fixed) transmit powers in the tiers if either (i) there is a single tier (see (4.74)), or (ii) the path-loss exponents are the same in all tiers and all tiers are open (see (5.60)).

If serving BS selection is made after applying selection bias (usually in order to modify the loading of the tiers in some way), we know that overall coverage is not the same as with the max-SINR selection criterion without bias (see Remark 5.6). In fact, some users may suffer from poor SINR due to interference from the tier that would have served them in the absence of selection bias, as discussed in Section 5.4. In these circumstances, operating the system at the desired level of performance calls for interference control, usually realized by means of *transmit power control* (or just power control, for short) with different levels of coordination across BSs, both in a tier and across tiers.

6.2 Power control from the transmitter perspective

Up until now, we have only considered links from the perspective of the receiver (the user in our case), because we are interested in the distribution of *received* SINR. However, *transmit* power control occurs at the transmitter, and it is therefore important to examine it from the perspective of the transmitters (BSs in our case). In particular, transmit power control can only occur on a link from a BS to a user if *that user is served by the BS*. When we refer to the distribution of received SINR at a user in a system using power control, we mean the distribution of received SINR at the user conditioned on its serving BS using power control on the link to this user, and on each interfering BS using power control on the link to *its* served user.

In short, the distribution of received SINR at an arbitrarily located user in the system is now a function of the distribution of transmit power at the BSs. Now, at any particular

BS, the distribution of transmit power by this BS depends on the number of users served by this BS, and the set of links to these served users. It follows that, in a system that uses power control, the distribution of SINR at any arbitrary user depends not only on the links to this user from the BSs in the system, but also on the links from these BSs to other users in the system that are served by these BSs. We shall see specific instances of such dependencies in later sections, but for now it suffices to realize that there are correlations between the links from a given BS to its set of served users. The fact that the links are correlated makes results like the Marking theorem (Theorem 3.4) inapplicable because the marks can no longer be treated as independent random variables. Thus the analysis in this chapter usually involves approximations such that we can continue to employ the PPP tools we used in Chapters 4 and 5.

We now begin our analysis of SINR distributions under power control by classifying the different power control schemes that have been deployed to date in terms of the extent of dependence of the distribution of transmit power from a given BS (to a served user) on the links to other users (either from the same BS, or from other BSs) in the system.

6.3 Types of power control

The most trivial type of power control is no power control, i.e. transmission with fixed power. This is the scenario that we have been analyzing so far. This kind of transmission mode can be generalized to one where the transmit power changes from one time period to another, but the values of transmitted powers are not dependent on the path losses or fade attenuations on any link in the system. In this book, we refer to this as *non-adaptive* power control. Note that, according to this definition, these variable transmit powers may even be random variables, so long as their distributions are independent of that of the path loss and fade attenuation on any link in the system. The LTE proposal for enhanced ICIC with ABS described in Section 5.4 is an example of such non-adaptive power control, because a predefined pattern of activity is specified for the macro BSs, independent of the quality of any link in the system. Further, if there is perfect synchronization across the macro BSs, then, at any instant, all macro BSs are transmitting with the same (fixed) power (which may be zero if this point in time is during an ABS). Thus the distribution of received SINR at the user can be calculated as the mixture of two distributions, one describing a non-ABS and the other an ABS instant in time.

The other general category of power control is *adaptive* power control, in which the transmit power is adapted to the conditions on one or more links in the system. We study two kinds of adaptive power control: open loop and closed loop.

(1) *Open-loop* power control (OLPC) Here the transmit power is a function of *only* the link from the transmitter to the intended destination, i.e. from the serving BS to the served user. For example, if the objective of transmit power control is to achieve a certain received power at the user, then OLPC effectively inverts the path loss on

the link to the user. However, because OLPC on a given link ignores the actions of all other simultaneously transmitting BSs, it cannot adapt to changes in interference power level. For this reason, OLPC is not implemented in cellular systems today. Nonetheless, OLPC is of interest as a mathematically tractable adaptive power control scheme, and we show examples of such analysis for practically relevant HCNs.

(2) *Closed-loop* power control (CLPC) This is the most general power control scheme, wherein the transmit power on each link takes into account the total interference on that link from all other simultaneously active links, with the objective of achieving some target SINR at the receiver. It is the version of power control implemented in cellular systems today. The operation of CLPC usually involves feedback from the receiver of the link (the user, in our case) to the transmitter (the BS, in our case) as to how close to the chosen SINR target the actual measured SINR on the link is, after which the transmitter then adjusts its transmit power for the next transmission, and so on. Despite the complexity of the power control scheme, some remarkably general results exist that not only prove the existence of a power allocation for all simultaneously active links, but also provide distributed algorithms that are guaranteed to converge to this power allocation on these links. Unfortunately, such results exist for only a finite number of links, so it is not known if they extend to the case where there are a countably infinite number of links, one from each BS modeled as a point of a PPP. Still, some analysis is possible under certain simplifying assumptions, as we show in Section 6.4.7.

We now derive the distribution of the received SINR at an arbitrarily located user in a multi-tier deployment when all BSs (including the serving BS for this user) employ the same kind of transmit power control (non-adaptive, open-loop, or closed-loop).

6.4 Distribution of SINR under power control

We begin our analysis with a basic result, which is another consequence of Theorem 4.1.

6.4.1 Distribution of received power with i.i.d. BS transmit powers

THEOREM 6.1 *(Madhusudhanan et al., 2011.) Consider a deployment of BSs whose locations are modeled by the points of a homogeneous PPP Ψ with density ν and an arbitrarily located user in this network. Suppose that the slope and intercept parameters of the path-loss model* (1.1) *describing the link from any BS to this user are α and K, respectively, and that the (transmit power, fade attenuation) pairs $\{(P_b^{\text{tx}}, H_b)\}_{b\in\Psi}$ on these links are i.i.d. with common joint PDF $f_{P^{\text{tx}},H}(\cdot,\cdot)$. Let (P^{tx}, H) be a random vector with this joint PDF. Then the one-dimensional point process defined by the received powers at the user from all BSs $b \in \Psi$,*

$$\tilde{\Psi} = \left\{ \frac{K P_b^{\text{tx}} H_b}{R_b^{\alpha}} : b \in \Psi \right\},$$

is equivalent to the one-dimensional point process of received powers at the users from all BSs of a different deployment, modeled by the points of a homogeneous PPP Φ *with density* $\lambda = \nu\,\mathbb{E}[(P^{\text{tx}}H)^{2/\alpha}]$*, all transmitting with fixed unit transmit power, with the same* α *and* K *for the path-loss model, and where the fade attenuations on all links from BSs in* Φ *to the user are i.i.d. with the same distribution as that of* $P^{\text{tx}}H$ *for the original deployment* Ψ*.*

Proof Just write the received power at the user from BS $b \in \Psi$ at distance R_b as

$$Y_b = \frac{KP_b^{\text{tx}}H_b}{R_b^\alpha} = \frac{K \times 1 \times \tilde{H}_b}{R_b^\alpha},$$

where $\tilde{H}_b \equiv P_b^{\text{tx}}H_b$, $b \in \Psi$. Then $\{\tilde{H}_b\}_{b\in\Psi}$ are i.i.d. with the same distribution as that of $\tilde{H} \equiv P^{\text{tx}}H$. The rest follows from Theorem 4.1. □

The consequence of Theorem 6.1 for our analysis is that, if the effect of power control at the BSs of each tier is that the transmit powers may be assumed i.i.d. (and independent of the fade attenuations on the links to the user, which themselves are assumed i.i.d.), then our existing results for fixed transmit powers at the BSs may be applied to analyze the distribution of the SINR at the user.

6.4.2 SINR distribution with non-adaptive power control

We assume throughout that all BSs in each tier are identical in their behavior, i.e. they employ the same power control scheme. If the power control scheme is non-adaptive, all BSs in a given tier do one of the following:

(1) use a random value of transmit power in each transmit interval that is drawn independently from a single common distribution (across all BSs), or
(2) repeatedly cycle through a single common (across all BSs) fixed finite-length sequence of transmit powers in successive transmit intervals.

The analysis of the first case follows immediately from Theorem 6.1, and lets us reuse the results derived in Section 5.3. This case also has some theoretical significance because, in a single-tier network of links where the transmitter locations are given by the points of a PPP, a random on–off non-adaptive transmit power pattern has been shown (Zhang & Haenggi, 2012) to yield a *Nash equilibrium*, meaning that no individual transmitter in the network can improve its throughput by transmitting using a different strategy.

For the second case, observe that, at any instant, the system has all BSs in the tier transmitting with the same fixed transmit power (that may change from one transmit interval to the next). Suppose the fixed sequence of transmit powers for BSs of tier i is $p_{i,0}^{\text{tx}}, \ldots, p_{i,N_i-1}^{\text{tx}}$ for N_i successive transmit intervals (assumed of equal duration), $i = 1, \ldots, n_{\text{tier}}$. In other words, at transmit interval t (counting from some reference time, corresponding to transmit interval 0), all BSs in tier i transmit with power $p_{i,t \bmod N_i}^{\text{tx}}$. Note that the same transmit power could be used in two different transmit intervals, i.e. $p_{i,0}^{\text{tx}}, \ldots, p_{i,N_i-1}^{\text{tx}}$ need not all be distinct.

As the lengths of the sequences $N_1,\ldots,N_{n_{\text{tier}}}$ could be different, we define N to be the least common multiple of $N_1,\ldots,N_{n_{\text{tier}}}$, then extend each of the sequences $p^{\text{tx}}_{i,0},\ldots,p^{\text{tx}}_{i,N_i-1}$ by repetition to obtain a sequence $p^{\text{tx}}_{i,0},\ldots,p^{\text{tx}}_{i,N-1}$ of length N. Then find all distinct n_{tier}-tuples $(p^{\text{tx}}_{1,j},\ldots,p^{\text{tx}}_{n_{\text{tier}},j})$ for $j\in\{0,1,\ldots,N-1\}$ and, label them $\boldsymbol{p}^{\text{tx}}_l$, $l=1,\ldots,N'$, where $N'\leq N$ and for each $l=1,\ldots,N'$ $\boldsymbol{p}^{\text{tx}}_l=[p^{\text{tx}}_{1,j(l)},\ldots,p^{\text{tx}}_{n_{\text{tier}},j(l)}]^\top$ for some $j(l)\in\{0,1,\ldots,N-1\}$.

For each $l=1,\ldots,N'$, define $\pi_l=n_l/N$, where n_l is the number of times $\boldsymbol{p}^{\text{tx}}_l$ appears in the set $\{(p^{\text{tx}}_{1,j},\ldots,p^{\text{tx}}_{n_{\text{tier}},j}) : j=0,1,\ldots,N-1\}$. Then we can say that the vector of (fixed) transmit powers by the BSs of the tiers $1,\ldots,n_{\text{tier}}$, $\boldsymbol{P}^{\text{tx}}\equiv[P^{\text{tx}}_1,\ldots,P^{\text{tx}}_{n_{\text{tier}}}]^\top$, takes the value $\boldsymbol{p}^{\text{tx}}_l$ with probability π_l, $l=1,\ldots,N'$, and, at any instant of time, the system is in one of these N' transmit-power scenarios with corresponding probabilities $\pi_1,\ldots,\pi_{N'}$.

It follows that the CCDF of the SINR at the user is a mixture of N' CCDFs, where the lth CCDF has mixture weight π_l and corresponds to the CCDF of the SINR (derived in Section 5.3) when the vector of transmit powers of the BSs in the tiers is given by $\boldsymbol{p}^{\text{tx}}_l$, $l=1,\ldots,N'$.

We now illustrate the application of these results to the case of eICIC in an LTE system.

6.4.3 Application: eICIC and feICIC in LTE

As has already been described earlier, eICIC is a scheme proposed in LTE macro-pico HCNs, where only the macro BSs employ power control, and this power control is of the second non-adaptive kind discussed in Section 6.4.2. To be precise, each macro BS transmits with fixed (maximum) power during a certain number of transmit intervals, then switches to fixed (near-zero) power during another number of transmit intervals (the ABSs), then repeats this cycle. The lower transmit power is not exactly zero because some essential control signaling always needs to be transmitted, and so the transmit power cannot fall below the minimum value required to do so. A more recent proposal aims to relax the requirement of near-zero power during ABSs and permit larger transmit powers during ABSs (though still lower than the transmit power during non-ABSs). This is called "further enhanced ICIC" or feICIC. We treat the more general feICIC case here, with eICIC being a special case. Note that this makes the ABSs *reduced-power* transmit intervals instead of (near) *zero-power* transmit intervals.

Label the macro tier by 1 and the pico tier by 2, both open, and assume there are no other tiers. Suppose that the fraction of ABSs is η_{ABS}, and that the transmit powers of macro BSs during ABSs and non-ABSs are $P^{\text{tx}}_{1,\text{red}}$ and $P^{\text{tx}}_{1,\text{reg}}$, respectively, where the transmit power in a "regular" transmit interval $P^{\text{tx}}_{1,\text{reg}}>P^{\text{tx}}_{1,\text{red}}$, the transmit power in a "reduced"-power transmit interval.

No time synchronization across the macro BSs

Suppose there is no time synchronization at all across the macro BSs. Consider an arbitrary macro BS in the system. At any given instant, this macro BS is in an ABS transmit interval with probability η_{ABS}, and in a non-ABS transmit interval with probability $1 - \eta_{\mathrm{ABS}}$, independently of any other macro BS. From Theorem 3.3, we conclude that, at any given instant, the macro tier Φ_1 with density λ_1 can be split into two independent tiers Φ_1' with density $\lambda_1' = \lambda_1(1 - \eta_{\mathrm{ABS}})$ and Φ_3 with density $\lambda_3 = \lambda_1 \eta_{\mathrm{ABS}}$, where all BSs in Φ_1' transmit with fixed power $P_{1,\mathrm{reg}}^{\mathrm{tx}}$, and all BSs in Φ_3 transmit with fixed power $P_{1,\mathrm{red}}^{\mathrm{tx}}$.

For eICIC, we can further simplify the problem by making the approximation that $P_{1,\mathrm{red}}^{\mathrm{tx}} = 0$, which means that the fraction of macros that have zero transmit power at any instant can be completely ignored in the analysis of SINR distribution at any UE. Thus, at a given instant, an arbitrary UE in the system is served by either the tier Φ_1' of non-zero-power macros (also called *active* macros), or the tier Φ_2 of pico BSs (with density λ_2, say), depending upon the serving tier selection criterion.

For simplicity, we assume that the candidate serving BS in each tier (Φ_1' or Φ_2) is the one nearest to the UE, and that the serving tier is chosen according to $I = \arg\min\{\tau_1 R_1^{\alpha}, \tau_2 R_2^{\alpha}\}$, where we assume the same path-loss exponent α on all links to the UE, and R_1 and R_2 are the distances to the nearest macro BS in Φ_1' and pico BS in Φ_2, respectively. Then the CCDF of the SINR at the UE is given by Theorem 5.10, with $n_{\mathrm{open}} = n_{\mathrm{tier}} = 2$, the two tiers being Φ_1' and Φ_2, with corresponding densities λ_1' and λ_2, and tier BS transmit powers fixed at $P_{1,\mathrm{reg}}^{\mathrm{tx}}$ and P_2^{tx}, respectively.

Perfect time synchronization across the macro BSs

The other extreme case is where there is perfect time synchronization across all the macro BSs, so, at any given instant, with probability $1 - \eta_{\mathrm{ABS}}$, all macro BSs are transmitting with power $P_{1,\mathrm{reg}}^{\mathrm{tx}}$ (because this instant falls in a non-ABS transmit interval), and with probability η_{ABS} all macro BSs are transmitting with power $P_{1,\mathrm{red}}^{\mathrm{tx}}$ (for feICIC). From the analysis of the second kind of non-adaptive power control in Section 6.4.2, we obtain the overall CCDF of the SINR at the UE to be

$$\mathbb{P}\{\Gamma_I > \gamma\} = (1 - \eta_{\mathrm{ABS}})\,\mathbb{P}\left\{\Gamma_I^{\mathrm{not\ ABS}} > \gamma\right\} + \eta_{\mathrm{ABS}}\,\mathbb{P}\left\{\Gamma_I^{\mathrm{ABS}} > \gamma\right\}, \quad \gamma > 0,$$

where the pico BSs always transmit with the same fixed power P_2^{tx}, and $\Gamma_I^{\mathrm{not\ ABS}}$ and Γ_I^{ABS} are given by Section 5.3 for the scenarios with all macro BS transmit powers fixed at $P_{1,\mathrm{reg}}^{\mathrm{tx}}$ and $P_{1,\mathrm{red}}^{\mathrm{tx}}$, respectively. We provide the details of the analysis in the context of spectral efficiency calculations in Section 7.2.3.

More realistic modeling assumptions

The results on the CCDF of the SINR derived in Section 5.3 do not distinguish between the non-ABS and ABS macro BS transmit power scenarios when the serving tier is that of the pico BSs. In other words, if the UE is served by a pico BS, the preceding analysis says that the pico BS will always transmit to this UE regardless of the transmit power of the interfering macro BSs. However, we know from the discussion in Section 5.4 that some UEs that are served by the pico tier (chosen after application of selection bias) could be strongly affected by interference from the macro tier during non-ABS transmit intervals, which affects the data rate to such UEs. Thus, in order to improve system throughput, it makes sense for the serving pico BS to *schedule* transmissions to such UEs in ABSs, when there is minimal interference from macro BSs. The topic of scheduling is a very complex one, and we cannot provide more than cursory treatment of it in this book. A brief overview of rate and throughput analysis for systems employing some widely used scheduling algorithms will be provided in Chapter 7.

Another implicit assumption made in the analysis in Section 5.3 is that the candidate serving BSs in the tiers are chosen independently of one another, i.e. the choice of a candidate serving pico BS, say, does not depend on the candidate serving macro BS. However, we have seen that the SINR can be improved by scheduling, and for scheduling to be successful we require synchronization between the candidate serving macro and pico BSs. This synchronization is easier to achieve in a practical deployment if the candidate serving macro and pico BSs for a given UE are physically near to one another. Now, such macro-pico HCNs are likely to be deployed as a pico-tier overlay on an existing macrocellular single-tier deployment, and since the macro BSs already have backhaul access, it is often easiest to route coordination signaling through them. In that case, each macro BS is responsible for coordination across itself and all pico BSs in the macrocell region where this macro BS is the candidate serving macro BS. This in turn means that the UE is not free to choose its candidate serving pico BS from the entire pico tier, but must restrict itself to the set of pico BSs that lie in the coordination region controlled by the candidate serving macro BS.

If the candidate serving macro BS is the one that is nearest to the UE location, a given macro BS is the candidate serving macro BS for all UEs that lie in the *Voronoi cell* (Haenggi *et al.*, 2009, Sec. IIC) with nucleus at this macro BS, where, for a given point pattern of nuclei in the plane, the Voronoi cell with nucleus at one of these nuclei points is defined as the two-dimensional region in the plane all of whose points are closer to this nucleus than to any other nucleus of the point pattern. In light of this discussion, a UE can no longer choose as its candidate serving pico BS the pico BS that is nearest to the UE, but must choose the nearest pico BS from among the pico BSs that lie in the Voronoi cell with nucleus given by the candidate serving macro BS for this UE. In other words, all probability calculations are to be performed by implicitly conditioning on the event that the candidate serving pico BS lies in this Voronoi cell. This has other implications on the analysis, as it changes the distribution of the distance of the candidate serving pico BS from the UE.

In Sections 6.4.4 and 6.4.5 we analyze the SINR distribution at an arbitrary user in a multi-tier HCN where the BSs in the tiers employ adaptive power control. As we shall see, there are some interesting nuances in the analysis that arise because transmit power control occurs at the transmitters, whereas the SINR is calculated at the receivers, so the analysis of SINR with power control involves a joint analysis of transmitters and receivers. We begin with the study of open-loop power control (OLPC).

6.4.4 Interference power at the receiver of a given link under OLPC

Under OLPC, each BS selects a target received power for the user it is serving, then transmits exactly as much power as is required to achieve this received power target.

Of course, to do this exactly would require exact instantaneous knowledge of the fade attenuation on the link from this BS to this user, which is usually impossible to obtain. Thus, OLPC usually means that the BS transmits with exactly as much power as required to achieve the received power target *in the absence of fading*. In other words, OLPC compensates for path loss but not for fading.

Consider a BS b in tier i (whose BSs are located at the points of a PPP Φ_i) that is the serving BS for some user at a distance of $R_{b,*}$ say. For future reference, we shall represent this user, served by the BS b, by u_b. Our interest will be in the link with transmitter b and receiver u_b, and specifically in the interference power at the receiver u_b of the given link.

Suppose the target received power for users receiving from BSs in tier i is set to ξ_i, $i = 1, \ldots, n_{\text{open}}$. Then BS b transmits with transmit power P_b^{tx} such that the received power in the absence of fading is ξ_i at this user:

$$\frac{K_i P_b^{\text{tx}}}{R_{b,*}^{\alpha_i}} = \xi_i \Leftrightarrow P_b^{\text{tx}} = \frac{\xi_i R_{b,*}^{\alpha_i}}{K_i}, \quad b \in \Phi_i, \quad i = 1, \ldots, n_{\text{open}},$$

where α_i and K_i are the slope and intercept, respectively, of the path-loss model for links from BSs in tier i to users. It follows that the received power at u_b from its serving BS b is given by

$$Y_b(u_b) = \frac{K_i P_b^{\text{tx}} H_{i,b}}{R_{b,*}^{\alpha_i}} = \xi_i H_{i,b},$$

where $H_{i,b}$ is the fade attenuation on the link to the user u_b from its serving BS $b \in \Phi_i$.

Now consider the total interference power at this user u_b. Let $\Phi = \bigcup_{i=1}^{n_{\text{tier}}} \Phi_i$ represent the total point pattern of BS locations in all tiers. Assume that BSs in the closed tiers $n_{\text{open}} + 1, \ldots, n_{\text{tier}}$ transmit with fixed transmit powers $P_{n_{\text{open}}+1}^{\text{tx}}, \ldots, P_{n_{\text{tier}}}^{\text{tx}}$, respectively, and that BSs in the open tiers $1, \ldots, n_{\text{open}}$ use OLPC on the links to their served users. Thus, any $b' \in \Phi \setminus \{b\}$ transmits to *its* served user $u_{b'}$ (at distance $R_{b',*}$ from b') with power

$$P_{b'}^{\text{tx}} = \frac{\xi_j R_{b',*}^{\alpha_j}}{K_j} \tag{6.1}$$

if $b' \in \Phi_j, j = 1, \ldots, n_{\text{open}}$. Let the distance of BS $b' \neq b$ from u_b be denoted $R_{b'}(u_b)$. Then the received power at the user of interest u_b from BS $b' \in \Phi_j$ is

$$Y_{b'}(u_b) = \frac{K_j P_{b'}^{\text{tx}} H_{j,b'}}{R_{b'}(u_b)^{\alpha_j}} = \xi_j \left(\frac{R_{b',*}}{R_{b'}(u_b)} \right)^{\alpha_j} H_{j,b'}, \quad b' \in \Phi_j, \quad j = 1, \ldots, n_{\text{open}}, \tag{6.2}$$

where $H_{j,b'}$ is the fade attenuation on the link between this interfering BS $b' \in \Phi_j$ and the user of interest u_b. Thus the total interference power at the user u_b is

$$W(u_b) = \sum_{b' \in \left(\bigcup_{i=1}^{n_{\text{open}}} \Phi_i\right) \setminus \{b\}} Y_{b'}(u_b) + \sum_{b'' \in \bigcup_{i=n_{\text{open}}+1}^{n_{\text{tier}}} \Phi_i} Y_{b''}(u_b),$$

where $Y_{b'}(u_b)$ is given by (6.2), and

$$Y_{b''}(u_b) = \frac{K_j P_j^{\text{tx}} H_{j,b''}}{R_{b''}(u_b)^{\alpha_j}}, \quad b'' \in \Phi_j, \quad j = n_{\text{open}} + 1, \ldots, n_{\text{tier}}.$$

Observe that, in each tier $j = 1, \ldots, n_{\text{open}}$, if the distance $R_{b',*}$ from each BS $b' \in \Phi_j$ to the user $u_{b'}$ served by it at that instant is i.i.d. across all $b' \in \Phi_j$, then $P_{b'}^{\text{tx}}$ is i.i.d. across all $b' \in \Phi_j$. It follows from Theorem 6.1 and (4.6) that, if the tiers have densities λ_j, $j = 1, \ldots, n_{\text{tier}}$, the one-dimensional point process of received powers at the user of interest u_b from all BSs b' from an open tier $j \in \{1, \ldots, n_{\text{open}}\} \setminus \{i\}$ is a PPP $\tilde{\Phi}_{j,u_b}$ with intensity function given by

$$\tilde{\lambda}_j(y; u_b) = \frac{2\pi\lambda_j \xi_j^{2/\alpha_j}\, \mathbb{E}[R_{j,*}^2]\, \mathbb{E}[H_j^{2/\alpha_j}]}{\alpha_j y^{2/\alpha_j + 1}}, \quad y \geq 0, \quad j \in \{1, \ldots, n_{\text{open}}\} \setminus \{i\}, \tag{6.3}$$

where $R_{j,*}$ is a random variable with the same distribution as that of $R_{b',*}$ for any $b' \in \Phi_j, j \neq i$, and H_j is a random variable with the same distribution as that of $H_{j,b'}$ for any $b' \in \Phi_j$. From Theorem 4.3, (6.3) also holds with $j = i$ for the intensity function of the PPP of received powers at u_b from all $b' \in \Phi_i \setminus \{b\}$, where $R_{i,*}$ is a random variable with the common distribution of $\{R_{b',*}\}_{b \in \Phi_i \setminus \{b\}}$.

Thus the Laplace transform of the total interference at user u_b from all BSs in tier $j, j = 1, \ldots, n_{\text{open}}$, can be obtained using (4.11) provided we know $\mathbb{E}[R_{j,*}^2]$. We now proceed to study the distribution of $R_{j,*}$.

6.4.5 Distribution of distance from BS to served user

Recall that $R_{j,*}$ is a random variable with the same distribution as that of the distance $R_{b',*}$ from each BS $b' \in \Phi_j$ to *its served user* $u_{b'}$ at the present instant. This is an important complication in the analysis, because we do not know which of the users for which this BS b' is the serving BS was actually *scheduled* by b' for transmission at this instant, so we do not know the location or identity of this selected served user $u_{b'}$. To be precise, let $\mathcal{U}_{b'}$ be the set of all users for which b' is the serving BS. Then, at each instant (transmit interval), the choice of which user $u_{b'}$ in $\mathcal{U}_{b'}$ is to be selected for transmission by b' is made by the scheduler, according to some scheduling policy that is yet to be discussed.

We emphasize here that $R_{b',*}$ is not the distance from BS b' to an *arbitrary* user location. Rather, it is the distance to a user $u_{b'}$ *selected in some way (by the scheduling policy) from the set* $\mathcal{U}_{b'}$ of users for which this BS b' is the serving BS. Owing to these additional conditions, the distribution of $R_{b',*}$ is not easy to determine even for the simplest scheduling policy and serving BS selection criteria, as we now show.

Distribution of distance from the nearest BS to an arbitrary served user

Consider a multi-tier network where the BS locations in each tier i are points of a homogeneous PPP Φ_i with density λ_i, $i = 1, \ldots, n_{\text{tier}}$. Suppose that, for each user,

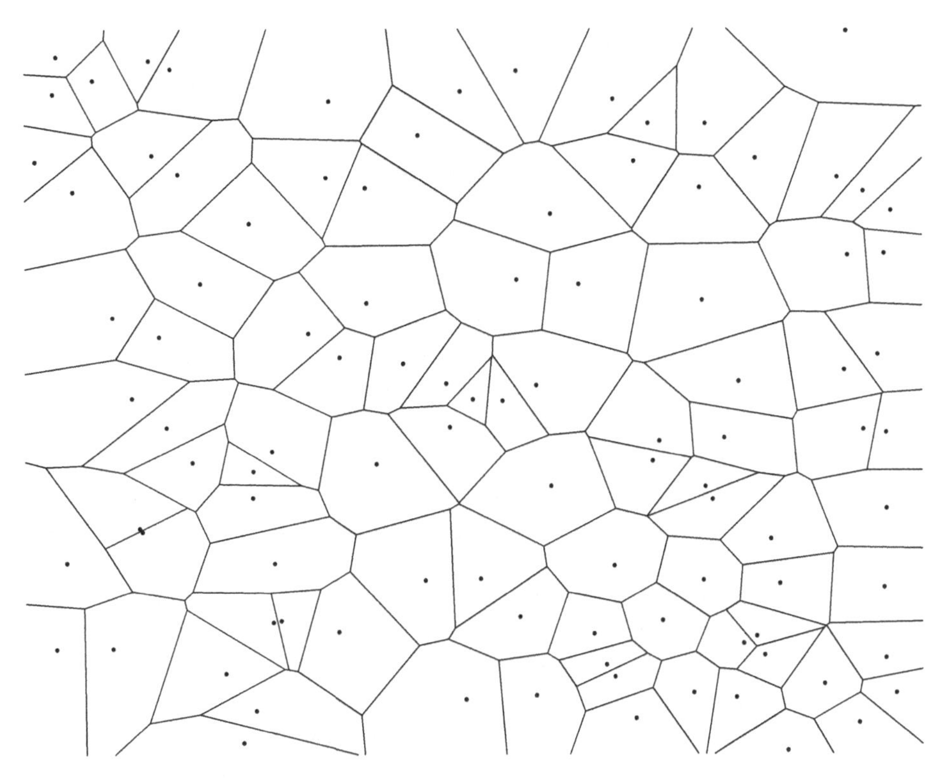

Figure 6.1 An example of (part of a) Voronoi tessellation induced by a homogeneous PPP modeling a single tier of BSs. Each polygonal region is a Voronoi cell. The dots represent the BSs, which are also the nuclei of the Voronoi cells. The boundaries of the cells are given by the perpendicular bisectors of the lines joining the nuclei.

the candidate serving BS from each open tier $i \in \{1, \ldots, n_{\text{open}}\}$ is the BS that is nearest to it in that tier. The spatial point pattern Φ_i of the BSs of each open tier i of the network creates a *Voronoi tessellation* (Calka, 2010) of the plane, i.e. a partition of the plane into a union of disjoint regions called *Voronoi cells*. Each Voronoi cell of the Voronoi tessellation formed from Φ_i contains exactly one BS from Φ_i, called the *nucleus* of this cell, and is defined as the region of the plane all of whose points are closer to this nucleus than to any other BS in the tier i, $i = 1, \ldots, n_{\text{open}}$. Figure 6.1 shows an example of the Voronoi tessellation induced by a single-tier homogeneous PPP.

If the serving tier for an arbitrarily located user is $I = i$, the serving BS for this user is just the nucleus of the Voronoi cell (of the Voronoi tessellation formed from Φ_i) in which this user lies. It follows that the distance from an arbitrary user to its serving BS (assumed to be in tier i) is just the distance from an arbitrary location in the network to the nearest point of the PPP Φ_i, and we already know that this distance has the PDF

$$2\pi\lambda_i r \exp\left(-\pi\lambda_i r^2\right), \quad r \geq 0. \tag{6.4}$$

The distribution of the distance from an arbitrary user to its serving BS has also been derived for multi-tier deployments and the serving BS selection criteria of Section 5.3.1 (see Jo *et al.*, (2012 Lem. 3)) and Theorem 5.12 (see Madhusudhanan *et al.*, (2012a, Lem. 4)).

However, for an arbitrary BS $b' \in \Phi_i$, the distance $R_{b',*}$ does not have the PDF (6.4) because, by definition, $R_{b',*}$ is the distance to a user $u_{b'}$ selected for transmission by this BS b' at this instant from among the users in the set $\mathcal{U}_{b'}$. Now, the set $\mathcal{U}_{b'}$ is the set of all users for which $b' \in \Phi_i$ is the serving BS, i.e. the set of all user locations for which the nucleus of the corresponding Voronoi cell (of the Voronoi tessellation formed from Φ_i) is b'. In other words, $\mathcal{U}_{b'}$ is just the set of all users that lie in the Voronoi cell (of the Voronoi tessellation formed from Φ_i) with nucleus b'.

Suppose the scheduler simply picks one of these users at random at every transmit interval and schedules a transmission to this user from $b' \in \Phi_i$. Then $R_{b',*}$ is the distance from b' to a randomly chosen user in the Voronoi cell (of the Voronoi tessellation formed from Φ_i) with nucleus b'. Note that $R_{b',*}$ does not have the PDF (6.4) (see also Dhillon, Ganti & Andrews (2013, Rem. 3)). Unfortunately, we do not know the distribution of $R_{b',*}$ because we do not know the distribution of the number of locations of the users in this Voronoi cell.

Let us therefore simplify the problem as much as possible and suppose that the user locations in the network are modeled by the points of another homogeneous PPP Φ_{u} with density λ_{u}, say, where Φ_{u} is independent of Φ_i, $i = 1, \ldots, n_{\mathrm{tier}}$. Then, conditioned on a given Voronoi cell of the Voronoi tessellation formed from Φ_i for some $i \in \{1, \ldots, n_{\mathrm{open}}\}$, the number of users lying within that Voronoi cell is a Poisson random variable, and, if we further condition on the number of these users in the Voronoi cell, then their locations are all i.i.d. uniformly distributed over the cell. The difficulty arises in calculating the unconditional distribution of the number of users in the Voronoi cell, and their locations within it.

The study of tessellations of the plane induced by random point patterns is well developed (Calka, 2010), with an extensive set of results for Voronoi cells induced by PPPs. In particular, if we take an arbitrary Voronoi cell from the tessellation created by the homogeneous PPP Φ_i of the locations of the BSs in tier i, then change the origin to be at the nucleus of this cell, we obtain the *typical cell* for the so-called *Poisson–Voronoi tessellation* (Calka, 2010, Sec. 5.1.2.4). The expected number of points of the independent PPP Φ_{u} that lie in this typical cell can be shown (Foss & Zuyev, 1996) to be $\lambda_{\mathrm{u}}/\lambda_i$ (see Problem 6.1).

In Yu & Kim (2013, Lem. 1), the distribution of the number N_{u} of users (with locations modeled by points of a homogeneous PPP with density λ_{u}) that lie in an arbitrary Voronoi cell created by an independent single-tier homogeneous PPP deployment of BSs with density λ is modeled by the following PMF:

$$\mathbb{P}\{N_{\mathrm{u}} = n\} = \frac{3.5^{3.5}\,\Gamma(n+3.5)}{n!\,\Gamma(3.5)} \frac{(\lambda_{\mathrm{u}}/\lambda)^n}{(\lambda_{\mathrm{u}}/\lambda + 3.5)^{n+3.5}}, \quad n = 0, 1, \ldots. \tag{6.5}$$

It can be shown (see Problem 6.2) that $\mathbb{E}[N_{\mathrm{u}}] = \lambda_{\mathrm{u}}/\lambda$, which matches the exact result.

However, the distribution of the distance $R_{b',*}$ from the nucleus of an arbitrary Voronoi cell to a randomly selected one of the N_u user locations inside this Voronoi cell (of the Voronoi tessellation formed from Φ_i) is still unkown, though some related results are known (Voss *et al.*, 2009). In Yu *et al.* (2012), the PDF of $R_{b',*}$ for $b' \in \Phi_i$ is approximated by

$$2\pi c\lambda_i r \exp\left(-\pi c\lambda_i r^2\right), \quad r \geq 0, \quad c = 1.25, \quad i = 1, \ldots, n_{\text{open}}. \tag{6.6}$$

In Figure 6.2, we show plots of the CDFs corresponding to the PDFs (6.4) and (6.6) together with empirical CDFs obtained via simulation for various choices of λ_u and $\lambda \equiv \lambda_1$ for a single tier ($n_{\text{open}} = n_{\text{tier}} = 1$) of pico BSs (Yu *et al.*, 2012). The plots indicate that the approximate PDF (6.6) is fairly accurate as long as λ is not "too small." The approximate PDF (6.6) has the benefit of being as tractable as (6.4). It is surprising that a single choice of $c = 1.25$ provides an adequate fit to a large range of values of λ, λ_u, and λ/λ_u, and that (6.6) does not depend on λ_u. There is some empirical justification for the form of (6.6) (see Gloaguen, Voss & Schmidt (2009)). In effect, the change in the value of the parameter c from 1 in (6.4) to 1.25 in (6.6) indicates that restricting the user locations to be inside a Voronoi cell has the effect of increasing the density of the BSs by 25%.

Remark 6.1 The empirical validity of (6.6) should be taken with caution because it has been established only for small cells and not macrocells, whereas our discussion in Section 5.4 indicates that transmit power control is more likely to be employed by the macro BSs than by the pico BSs.

Distribution of distance from strongest BS to an arbitrary served user

Recall from Lemma 4.12 that, for a single tier, the distance from an arbitrary user to the strongest BS in that tier (i.e. the candidate serving BS from that tier) has PDF (4.49). Since this is of the same form as (6.4) and the PPP of received powers at the user from the BSs of that tier with arbitrary i.i.d. fading is equivalent to the PPP of received powers from a different BS deployment for that tier but with no fading, it should be possible to extend the approximation in the preceding subsection and say that the PDF of the distance from an arbitrary BS in the tier i to a user, randomly selected from the set of users for which this BS is the serving BS, is approximately given by

$$2\pi c\lambda_i \mathbb{E}[H_i^{2/\alpha}] r \exp\left(-\pi c\lambda_i \mathbb{E}[H_i^{2/\alpha}] r^2\right), \quad r \geq 0, \quad i = 1, \ldots, n_{\text{open}},$$

where H_i is a random variable with the common distribution of the fade attenuations on the links from all BSs in tier i to the user. We do not pursue this any further in this book, and will restrict ourselves to selecting candidate serving BSs from the tiers as the nearest BSs to the user of interest.

Problems

6.1 Prove that the expected number of points of a homogeneous PPP Φ_u with density λ_u that lie in the typical cell $C(\Phi)$ of the Poisson–Voronoi tessellation created by the

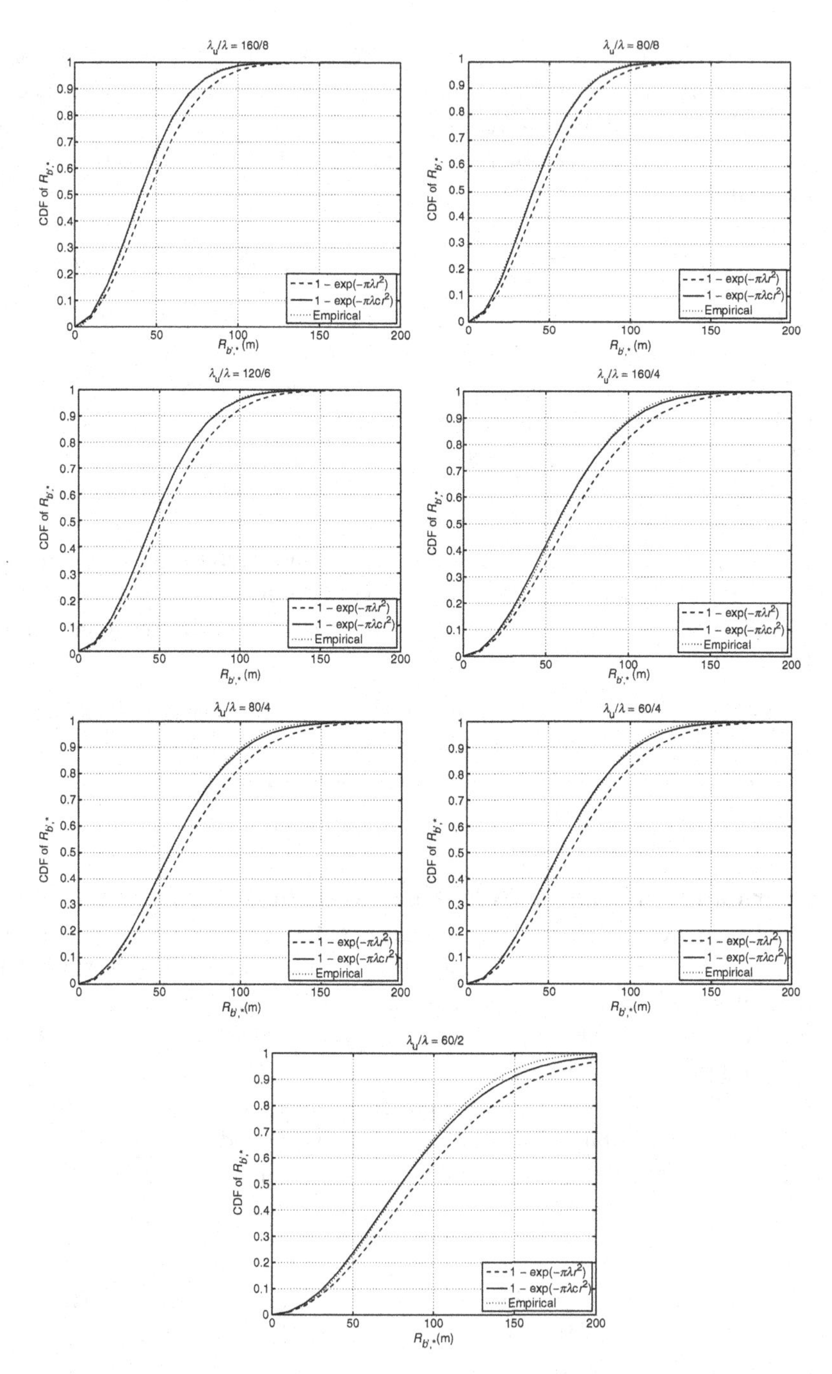

Figure 6.2 Plots of empirical CDF for distance $R_{b',*}$ between an arbitrary pico BS b' and one of the users in $\mathcal{U}_{b'}$, chosen at random, obtained via simulation on a 13-macrocell region with wraparound and inter-macrosite distance of 500 m. The various plots are for different choices of $\lambda_\mathrm{u}/\lambda = m/n$, which is interpreted as m users and n BSs per 120 deg-sector of each of the 13 macrocells. It is seen to be accurately approximated by the CDF corresponding to the PDF (6.6), except when λ becomes small.

homogeneous PPP Φ with density λ is given by λ_u/λ. *Hint* (Foss & Zuyev, 1996): The desired quantity is

$$\mathbb{E} \sum_{(x,y)\in\Phi_u} 1_{C(\Phi)}(x,y) = \mathbb{E}\left\{\mathbb{E}\left[\sum_{(x,y)\in\Phi_u} 1_{C(\Phi)}(x,y) \mid \Phi\right]\right\}$$

$$= \lambda_u \mathbb{E}\left[\int_{-\infty}^{\infty}\int_{-\infty}^{\infty} \mathbb{P}\{(x,y) \in C(\Phi)\}\, dx\, dy \mid \Phi\right] \quad (6.7)$$

$$= \lambda_u \mathbb{E}\left\{\int_{-\infty}^{\infty}\int_{-\infty}^{\infty} \exp[-\pi\lambda(x^2+y^2)]\, dx\, dy \mid \Phi\right\}, \quad (6.8)$$

where (6.7) follows from (4.63) and in (6.8) we use the fact that $C(\Phi)$ has the nucleus at the origin, so a point $(x,y) \in C(\Phi)$ if and only if the origin is the nearest point of the PPP Φ to the location (x,y), i.e. there is no point of Φ in the open disk of radius $\sqrt{x^2+y^2}$ centered at (x,y). Switching to polar coordinates in (6.8) yields the desired result.

6.2 Verify that for N_u with PMF given by (6.5), $\mathbb{E}[N_u] = \lambda_u/\lambda$. *Hint*: Write (6.5) as

$$\mathbb{P}\{N_u = n\} = \frac{\Gamma(n+3.5)}{n!\,\Gamma(3.5)}\left(\frac{\lambda_u/\lambda}{\lambda_u/\lambda+3.5}\right)^n\left(1-\frac{\lambda_u/\lambda}{\lambda_u/\lambda+3.5}\right)^{3.5}$$

$$= \frac{\Gamma(n+r)}{n!\,\Gamma(r)}p^n(1-p)^r, \quad n = 0, 1, \ldots,$$

which is recognized as the generalized negative binomial (see Appendix A) or Pólya distribution with parameters $r = 3.5$ and $p = (\lambda_u/\lambda)/(\lambda_u/\lambda + 3.5)$, and is known to have mean $pr/(1-p)$.

6.4.6 CCDF of SINR when all BSs use OLPC

In this section, we consider an arbitrary user in the network. For each open tier, the candidate serving BS from that tier is chosen as the nearest BS from that tier to the user location. We begin with the analysis for a single-tier network, where the nearest BS to the user location is also the serving BS.

Single-tier network

Consider a single-tier deployment of BSs whose locations are modeled by a homogeneous PPP Φ with density λ, and users whose locations are modeled by an independent homogeneous PPP Φ_u with density λ_u. Let the fade attenuations on all links from all BSs to all users be i.i.d., and let H be a random variable with this common distribution. Suppose that the serving BS for each user is the BS that is nearest to the user. Consider an arbitrary user in the system and denote it u_b, where $b \in \Phi$ is its serving (i.e. nearest) BS. Let the distance from u_b to its serving BS b be denoted R_1, as shown in Figure 6.3. Then we know that R_1 has the PDF (6.4). By hypothesis, b transmits with OLPC to the served user u_b, so that the received power at u_b is ξH_b, say, where H_b is a random variable with the same distribution as that of H.

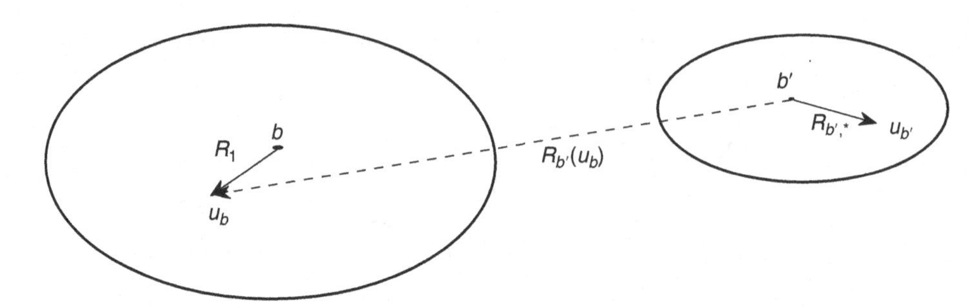

Figure 6.3 A link of interest in a single-tier network where all BSs transmit with open-loop power control (OLPC).

At the same time, each other BS $b' \neq b$ transmits to one of *its* served users selected at random, denoted $u_{b'}$, and all these other BSs also employ OLPC. From (6.2), the interference at user u_b from a BS $b' \neq b$ is given by

$$Y_{b'}(u_b) = \xi \left[\frac{R_{b',*}}{R_{b'}(u_b)}\right]^\alpha H_{b'} = \frac{\xi \tilde{H}_{b'}}{[R_{b'}(u_b)]^\alpha}, \quad b' \in \Phi \setminus \{b\},$$

where $H_{b'}$ for $b' \neq b$ are all i.i.d. with the same distribution as H, $R_{b'}(u_b)$ is the distance of BS b' from the user u_b, $R_{b',*}$ is the distance from the BS b' of the (unknown, selected at random from $\mathcal{U}_{b'}$) user $u_{b'}$ served by BS b' at this time instant (with PDF given by (6.6)), and $\tilde{H}_{b'} = R^\alpha_{b',*} H_{b'}$, $b' \in \Phi \setminus \{b\}$. Thus $\tilde{H}_{b'}$ are i.i.d. for all $b' \in \Phi \setminus \{b\}$. Let R_* be a random variable with the common PDF (6.6) of the i.i.d. random variables $R_{b',*}$, $b' \in \Phi \setminus \{b\}$. Then $\tilde{H} = HR^\alpha_*$ is a random variable with the common distribution of the i.i.d. random variables $\tilde{H}_{b'}$, $b' \in \Phi \setminus \{b\}$.

Conditioned on $R_1 = r_1$, we see that the total interference from all other BSs $b' \neq b$ at the user of interest u_b is given by

$$V(u_b) = \sum_{b' \in \Phi \setminus \{b\}} Y_{b'}(u_b) = \sum_{b' \in \Phi} \frac{\xi \tilde{H}_{b'}}{[R_{b'}(u_b)]^\alpha} 1_{(r_1,\infty)}(R_{b'}).$$

From (4.24) and (4.25) we obtain the Laplace transform of $V(u_b)$, conditioned on the distance from user u_b to its serving BS b being $R_1 = r_1$, as follows:

$$\mathcal{L}_{V(u_b)\,|\,R_1=r_1}(s) = \exp\left[-\pi\lambda(\xi s)^{2/\alpha}\tilde{G}_\alpha\left(\frac{r_1^2}{\xi^{2/\alpha}s^{2/\alpha}}\right)\right], \quad s > 0,$$

$$\begin{aligned}\tilde{G}_\alpha(z) &= \mathbb{E}\left[\gamma\left(1-\frac{2}{\alpha}, \frac{\tilde{H}}{z^{\alpha/2}}\right)\tilde{H}^{2/\alpha}\right] - z\left[1 - \mathcal{L}_{\tilde{H}}\left(\frac{1}{z^{\alpha/2}}\right)\right] \\ &= \mathbb{E}\left[\gamma\left(1-\frac{2}{\alpha}, \frac{HR_*^\alpha}{z^{\alpha/2}}\right)R_*^2 H^{2/\alpha}\right] - z\left\{1 - \mathbb{E}\left[\mathcal{L}_H\left(\frac{R_*^\alpha}{z^{\alpha/2}}\right)\right]\right\}, \quad z \geq 0.\end{aligned}$$

The CCDF of the SINR at the user of interest u_b, both conditioned on $R_1 = r_1$ and unconditional, can then be obtained using the same steps as in the derivations of (4.33) and (4.38), respectively.

Multi-tier network, serving BS is "nearest" (after selection bias)
Consider an arbitrarily located user. The candidate serving BSs from the open tiers are the ones nearest to the user location. If their distances from the user are denoted $R_1, \ldots, R_{n_{\text{open}}}$, the serving tier is $I = \arg\min_{1 \le i \le n_{\text{open}}} \tau_i R_i^{\alpha_i}$, and the serving BS is the candidate serving BS from tier I, denoted B_I. Using the notation given previously, the user is now denoted u_{B_I}. The SINR at u_{B_I} when receiving from its serving BS B_I under OLPC (and when all other BSs are also transmitting using OLPC to whichever of their served users they have selected to transmit to at this instant) is given by

$$\Gamma_I(u_{B_I}) = \frac{\xi_I H_{I,*}}{\sum_{b' \in \Phi_I \setminus \{B_I\}} \frac{\xi_I \tilde{H}_I}{[R_{b'}(u_{B_I})]^{\alpha_I}} + \sum_{\substack{j=1 \\ j \ne I}}^{n_{\text{open}}} \sum_{b' \in \Phi_j} \frac{\xi_j \tilde{H}_{j,b'}}{[R'_b(u_{B_I})]^{\alpha_j}} + \sum_{l=n_{\text{open}}+1}^{n_{\text{tier}}} \sum_{b'' \in \Phi_l} \frac{P_l H_{l,b''}}{[R_{b''}(u_{B_I})]^{\alpha_l}}}.$$

The CCDF of the SINR is given by

$$\begin{aligned}
\mathbb{P}\left\{\Gamma_I(u_{B_I}) > \gamma\right\} &= \sum_{i=1}^{n_{\text{open}}} \mathbb{P}\left\{ i = \arg\min_{1 \le j \le n_{\text{open}}} \tau_j R_j^{\alpha_j},\ \Gamma_i(u_{B_i}) > \gamma \right\} \\
&= \sum_{i=1}^{n_{\text{open}}} \mathbb{P}\left\{ i = \arg\min_{1 \le j \le n_{\text{open}}} \tau_j R_j^{\alpha_j},\ \frac{\xi_i H_{i,*}}{\sum_{\substack{j=1 \\ j \ne i}}^{n_{\text{open}}} \frac{\xi_j \tilde{H}_j}{R_j^{\alpha_j}} + W(R_1, \ldots, R_{n_{\text{open}}}; u_{B_i})} > \gamma \right\},
\end{aligned} \tag{6.9}$$

where $\tilde{H}_j \equiv R_{B_j,*}^{\alpha_j} H_{j,B_j}$, $j \in \{1, \ldots, n_{\text{open}}\} \setminus \{i\}$, and, for any BS b and served user u_b, we define

$$\begin{aligned}
W(r_1, \ldots, r_{n_{\text{open}}}; u_b) &= \sum_{j=1}^{n_{\text{open}}} V_j(r_j; u_b) + \sum_{l=n_{\text{open}}+1}^{n_{\text{tier}}} W_l(u_b) + N_0, \\
V_j(r; u_b) &= \sum_{b' \in \Phi_j} \frac{\xi_j \tilde{H}_{j,b'}}{[R_{b'}(u_b)]^{\alpha_j}} 1_{(r,\infty)}(R_{b'}(u_b)), \quad r > 0, \quad j = 1, \ldots, n_{\text{open}}, \\
W_l(u_b) &= \sum_{b'' \in \Phi_l} \frac{P_l H_{l,b''}}{[R_{b''}(u_b)]^{\alpha_l}}, \quad l = n_{\text{open}} + 1, \ldots, n_{\text{tier}}.
\end{aligned}$$

We see that the right side of (6.9) can be calculated from Theorem 5.10, with the change that $P_1, \ldots, P_{n_{\text{open}}}$ are replaced by $\xi_1, \ldots, \xi_{n_{\text{open}}}$, respectively, and the fade attenuations to the user from all BSs (other than the serving BS) in an open tier $j \in \{1, \ldots, n_{\text{open}}\}$ are i.i.d. with the same distribution as the random variable $\tilde{H}_j$.

Multi-tier network, serving BS is "strongest" (after selection bias)

Recall the discussion leading up to (5.71), where we distinguish between the pilot and traffic channels. Note that the transmit power on the pilot channel is always fixed. For fixed transmit power on the traffic channel, the analysis in Section 5.3.2 assumed that the serving tier selection was followed immediately by transmission, so that the fade attenuations on the pilot and traffic channels could be assumed identical. However, if the transmit power is determined from OLPC, there is likely to be a time interval between serving tier selection and transmission from the serving BS because of the additional processing needed to perform OLPC. Thus, the scenario in Remark 5.9 is the more likely one when we have OLPC, and we can assume that, for each BS b in each open tier i, the fade attenuations on the pilot and traffic channels on the link to any given user, $H_{i,b}^{\text{pilot}}$ and $H_{i,b}^{\text{traffic}}$, are i.i.d. but not identical. We can write the CCDF of the SINR on the traffic channel at the user as follows:

$$\mathbb{P}\left\{\Gamma_I^{\text{traffic}} > \gamma\right\} = \sum_{i=1}^{n_{\text{open}}} \mathbb{P}\left\{ i = \arg \max_{1 \le j \le n_{\text{open}}} \frac{\tau_j K_j P_j^{\text{pilot,tx}} H_j^{\text{pilot}}}{R_j^{\alpha_j}}, \right. \left. \frac{\xi_i H_{i,*}^{\text{traffic}}}{\sum_{\substack{j=1 \\ j \neq i}}^{n_{\text{open}}} \frac{\xi_j \tilde{H}_j^{\text{traffic}}}{R_j^{\alpha_j}} + W(R_1, \ldots, R_{n_{\text{open}}}; u_{B_i})} > \gamma \right\}. \tag{6.10}$$

If we condition on $\boldsymbol{R} \equiv [R_1, \ldots, R_{n_{\text{open}}}]^\top = \boldsymbol{r} \equiv [r_1, \ldots, r_{n_{\text{open}}}]^\top$, say, then, from the given assumption of independence of the fade attenuations, we can write

$$\mathbb{P}\{\Gamma_I^{\text{traffic}} > \gamma\} = \sum_{i=1}^{n_{\text{open}}} \int_{\mathbb{R}_+^{n_{\text{open}}}} \mathbb{P}\left\{ i = \arg \max_{1 \le j \le n_{\text{open}}} \frac{\tau_j K_j P_j^{\text{pilot,tx}} H_j^{\text{pilot}}}{r_j^{\alpha_j}} \right\} \times \mathbb{P}\left\{ \frac{\xi_i H_{i,*}^{\text{traffic}}}{\sum_{\substack{j=1 \\ j \neq i}}^{n_{\text{open}}} \frac{\xi_j \tilde{H}_j^{\text{traffic}}}{r_j^{\alpha_j}} + W(r_1, \ldots, r_{n_{\text{open}}}; u_{B_i})} > \gamma \right\} f_{\boldsymbol{R}}(\boldsymbol{r})\, \mathrm{d}\boldsymbol{r}.$$

Each term in the integral can be calculated using our previous results. For the second term, in order to apply the results of Section 5.2.1 we need to determine the distribution of $\tilde{H}_{j,b'}^{\text{traffic}} \equiv H_{j,b'}^{\text{traffic}} R_{b',*}^{\alpha_j}$, for all $j = 1, \ldots, n_{\text{open}}$ and for all BSs b' other than the serving BS. This can be achieved using the PDF (6.6) for $R_{b',*}^{\alpha_j}$.

On the other hand, if we assume that transmission from the selected serving tier occurs instantaneously after serving tier selection, so that $H_{i,b}^{\text{pilot}} \equiv H_{i,b}^{\text{traffic}} \equiv H_{i,b}$, say, then (6.10) can be calculated directly after conditioning on $\boldsymbol{R}$ using results like (5.79) and Theorem 5.11. However, to apply (5.79) after conditioning on $\boldsymbol{R}$, we need to again condition on the distances $R_{j,*}$ from the non-serving candidate serving BSs to *their* selected served users, so that (6.10) can be written in terms of $H_{1,*}, \ldots, H_{n_{\text{open}},*}$. We will not pursue this further in this book.

6.4.7 SINR distribution under CLPC

The subject of power control is deep and rich, and in this book we cannot provide even a cursory overview of the topic or principal results. Ideally, CLPC compensates not only for the path loss (as OLPC does), but also for the fade attenuation *and interference*, so that a target received SINR value can be achieved on a collection of links. Clearly, the transmit power on each link is now dependent on the transmit power on every other link. For a finite collection of interfering links, a remarkable result (Foschini & Miljanic, 1993) shows that, subject to a certain condition on the matrix of normalized path losses between every transmitter and every receiver in the system, not only does there exist a power allocation such that all target link SINRs can be achieved, but also that this optimal power allocation can be obtained using a simple iterative distributed algorithm that is guaranteed to converge. Unfortunately, our modeling of the network has a countably infinite number of links, and there are no results comparable to Foschini & Miljanic (1993) for this case. It is not even known whether the distributed algorithm in Foschini & Miljanic (1993) will converge in such a scenario, even if there exists a power allocation achieving the desired target SINRs on all links.

If a transmit power allocation does exist such that all links achieve their desired SINR targets, then of course there is no longer any SINR "distribution" because the SINR at any user is exactly the target SINR with probability 1. Consider a single-tier network where the BS locations are modeled by a homogeneous PPP Φ with density λ. If we assume that (a) such a transmit power allocation does exist, (b) all SINR targets are equal (to γ, say), and (c) the transmit powers at the BSs are i.i.d., it is possible to obtain this common distribution of these BS transmit powers as follows:

$$\mathbb{P}\{P^{\text{tx}} \leq p\} = \mathbb{P}\{\Gamma_I(p) \geq \gamma\}, \tag{6.11}$$

where $\Gamma_I(p)$ is the received power at an arbitrary user in the network from its serving BS when the latter transmits with power p, all other BSs transmitting with powers drawn i.i.d. from the same distribution. From the results in Section 5.3 and Theorem 6.1, we note that $\mathbb{P}\{\Gamma_I(p) \geq \gamma\}$ is a function of $\mathbb{E}[(P^{\text{tx}})^{2/\alpha}]$, but this is unknown unless we know the distribution of P^{tx}. Substituting the expression for $\mathbb{P}\{\Gamma_I(p) \geq \gamma\}$ in terms of $\mathbb{E}[(P^{\text{tx}})^{2/\alpha}]$ into (6.11), then using the resulting expression for the distribution of P^{tx} to write the expression for $\mathbb{E}[(P^{\text{tx}})^{2/\alpha}]$, we obtain a fixed-point equation for $\mathbb{E}[(P^{\text{tx}})^{2/\alpha}]$ that may be solved numerically to yield an expression for $\mathbb{P}\{P^{\text{tx}} \leq p\}$.

Note that we cannot ignore thermal noise power, because we already know that, for an interference-limited system, the distribution of SINR at an arbitrary user in the network does not depend on the transmit powers at the BSs (so long as they are i.i.d.). The presence of thermal noise may, however, also require unbounded transmit powers in order to achieve the desired link target SINR.

If we impose the physically realistic requirement of a certain maximum transmit power $p_{\max}$ for each BS, there will be a certain fraction of links where the target SINR cannot be achieved even when the BS transmits with maximum power. In other words, a fraction $q = \mathbb{P}\{P^{\text{tx}} > p_{\max}\}$ of the BSs will be transmitting with maximum power $p_{\max}$, while the rest will transmit with i.i.d. transmit powers $\tilde{P}^{\text{tx}}$ with common distribution given by that of P^{tx} truncated at $p_{\max}$:

$$\mathbb{P}\{\tilde{P}^{\text{tx}} \leq p\} = \frac{\mathbb{P}\{P^{\text{tx}} \leq p\}}{\mathbb{P}\{P^{\text{tx}} \leq p_{\max}\}}, \quad 0 \leq p \leq p_{\max}.$$

By partitioning the network Φ into (i) a PPP with density λq obtained by thinning Φ with retention probability q, where all transmit powers of BSs in this new PPP are fixed at $p_{\max}$, and (ii) a PPP with density $\lambda(1 - q)$, where all BSs in this PPP transmit with i.i.d. powers with the above distribution of $\tilde{P}^{\text{tx}}$, we can apply the existing body of results to determine the distribution of SINR at an arbitrary user that is served by one of the BSs with transmit power $p_{\max}$ (Erturk *et al.*, 2013).

7 Spectral and energy efficiency analysis

7.1 Introduction

We have so far studied only the distribution of the SINR at an arbitrarily located user in a single-tier cellular network or a multi-tier HCN. The CCDF of the SINR directly yields a measure of system performance called *coverage*, as discussed in Section 5.2.4. However, of equal and possibly greater interest to the network designer is another metric variously called *throughput*, *spectral efficiency*, and *capacity*. We favor the use of the term *spectral efficiency*.

The spectral efficiency can be calculated for each tier, and is defined as the total number of bits of information that an arbitrary BS in that tier can transfer to the set of its served users per use of the wireless "channel" to the users, i.e. per second and per hertz of bandwidth. We shall see that, because the users served by each BS are distributed over some region (the "cell"), this measure of spectral efficiency is an *area-averaged* quantity, and we shall sometimes refer to it as *area-averaged spectral efficiency* in order to emphasize this. Further, it is dependent on the details of the scheduling algorithm employed by the BS to select served users for transmission at each transmit interval. However, for the simplest scheduling scheme, where each of the served users is selected in turn for the same number of transmit intervals (called *round robin* or RR scheduling), we shall show that the area-averaged spectral efficiency is the same as the spectral efficiency on the link to a *single* randomly selected user served by the BS. Since the latter spectral efficiency on a single link is related to the received SINR on that link by the Shannon formula, the results derived in the previous chapters can be applied to derive the area-averaged spectral efficiency under RR scheduling. We also define the spectral efficiency of the entire network and show how to calculate it. A comparison of spectral efficiency of a single-tier dense small-cell network versus a two-tier macro-pico co-channel HCN quantifies the benefits of small-cell deployments in future wireless networks.

Another metric that has lately gained in importance is the *energy efficiency* of the network, which is calculated per tier and is the ratio of spectral efficiency to mean consumed power for an arbitrary BS in that tier. The energy efficiency of the entire network can also be calculated using our results for spectral efficiency, provided we have a means of computing the total power consumption (not just power transmitted over the air). We will not cover this in detail, but we briefly mention a widely used model for power consumption in a wireless cellular network.

We begin with the definition of spectral efficiency.

7.2 Spectral efficiency

7.2.1 Spectral efficiency on the link to an arbitrarily located user

We first recall the relationship between the spectral efficiency on the link to a single arbitrarily located user and the SINR on that link. Also we recall that all the analysis in this book has been for a "snapshot" of a network deployment, i.e. at one instant in time. For a given set of BS locations, this analysis may also be assumed to apply over a time duration (T, say) short enough that the fades on the various links may be assumed not to vary over that duration.

Consider a (possibly multi-tier) network Φ at a single instant in time, say the beginning of a transmit interval. Consider an arbitrarily located user. Let us assume that the duration of a transmit interval is short compared to T, so that, over the transmit interval, the distances $\{R_b\}_{b\in\Phi}$ of all BSs in all tiers may be assumed fixed, as also are the fade attenuations $\{H_b\}_{b\in\Phi}$ on all links to this user from all BSs in the system. Conditioned on this set of distances and fade attenuations, it can be shown under a minimal set of assumptions that the complex equivalent baseband signal received at the user is a circularly symmetric complex Gaussian random variable (Pinto & Win, 2010b, eqn. (3) and fn. 5).[1]

In the terminology of information theory, a single transmission made to this user from its serving BS during a transmit interval counts as a single "use" of the wireless "channel" between them. It follows that, *conditioned on the distances and fade attenuations* $\{(R_b, H_b)\}_{b\in\Phi}$, the spectral efficiency C on the link to this arbitrarily located user from its serving BS (in the sense of maximum mutual information that can be transmitted from the serving BS to this user per use of the wireless "channel" between them) is given by the Shannon formula (Pinto & Win, 2010b, eqn. (7)):

$$C = \log_2(1 + \Gamma_I) \text{ bits/s/Hz}, \tag{7.1}$$

where Γ_I is the SINR at this user (i.e. the SINR when receiving from its serving BS).

Remark 7.1 It follows from (7.1) that, for any BS, the (unconditional) capacity on the link to an arbitrarily located user in the network is a random variable, just like the SINR on that link. Thus, an outage probability may be defined for capacity in the same way as an outage (i.e. complement of coverage) probability is defined for the SINR, as the probability that the capacity on the link to an arbitrarily located user in the network is below some threshold, x bits/s/Hz, say. In fact, the capacity outage probability may be obtained from the SINR distribution as follows:

$$\mathbb{P}\{C \le x\} = \mathbb{P}\{\Gamma_I \le 2^x - 1\}, \quad x > 0.$$

[1] This implicitly assumes that the BSs transmit symbols from a complex Gaussian "alphabet." This alphabet and thermal noise on the links to the user are the only sources of randomness in the system because the power attenuation on each link is fixed, given the set of distances and fade attenuations.

Owing to this one-to-one relationship between capacity outage and SINR outage, we do not consider capacity outage further in this book.

Note that (7.1) is the spectral efficiency on the link to an arbitrarily located user conditioned on the set of distances and fade attenuations on all links to the user from all BSs. Thus the unconditional spectral efficiency μ_C on the link to this user is the expectation over the point pattern of BS locations, and over the fade attenuations on all links to the user from these BSs:

$$\mu_C \equiv \mathbb{E}[C] = \int_0^\infty \mathbb{P}\{C > x\}\mathrm{d}x \tag{7.2}$$

$$= \int_0^\infty \mathbb{P}\{\Gamma_I > 2^x - 1\}\mathrm{d}x = \frac{1}{\ln 2}\int_0^\infty \frac{1}{1+\gamma}\mathbb{P}\{\Gamma_I > \gamma\}\mathrm{d}\gamma, \tag{7.3}$$

where (7.2) follows from the identity

$$\mathbb{E}[X] = \int_0^\infty \mathbb{P}\{X > x\}\mathrm{d}x \tag{7.4}$$

for any positive-valued random variable X (see Problem 7.1), and in (7.3) we have made the change of variable of integration from x to $\gamma = 2^x - 1$. Then we can use the expressions for $\mathbb{P}\{\Gamma_I > \gamma\}$ derived in Sections 5.3 and 6.4 in (7.3) to evaluate the spectral efficiency on the link to an arbitrarily located user.

Remark 7.2 For the case of a single-tier network, no thermal noise, i.i.d. Rayleigh fading on all links to the user, path-loss exponent $\alpha = 4$, and the user being served by the nearest BS, we know from (4.37) that

$$\mathbb{P}\{\Gamma > \gamma\} = \frac{1}{1 + \sqrt{\gamma}\cot^{-1}(1/\sqrt{\gamma})}, \quad \gamma > 0,$$

so from (7.3) we see that the spectral efficiency of the link to an arbitrarily located user in this system is

$$\mu_C = \frac{2}{\ln 2}\int_0^{\pi/2} \frac{\tan\theta}{1 + \theta\tan\theta}\mathrm{d}\theta = \frac{1.49}{\ln 2}\text{ bits/s/Hz}, \tag{7.5}$$

where we have made the change of variables from γ to $\theta = \cot^{-1}(1/\sqrt{\gamma})$. This result was first proved in Andrews *et al.* (2011, eqn. (17)).

Spectral efficiency as area-averaged rate

For a snapshot of the system, the expressions derived in Sections 5.3 and 6.4 are the *area-averaged* distributions of the SINR under various association criteria. In other words, if we were to sample a single instance of the BS deployment at a very large number of locations, this would be the empirical CDF of the SINRs at those locations. Therefore $\mathbb{P}\{C > x\}$ for $x > 0$ is the *area-averaged* distribution of the achievable rates to users under the corresponding association criteria. In particular, $\mu_C = \mathbb{E}[C]$ is the area-averaged *mean downlink rate*.

Relation to ergodic rate

It is important to understand, however, that, notwithstanding its resemblance to the formula for *ergodic rate*, $\mu_C = \mathbb{E}[C] = \mathbb{E}[\log_2(1 + \Gamma_I)]$ is *not* the ergodic rate to any single user. The reason is that the definition of ergodic rate assumes that transmissions occur over such a long time as to encounter every possible state of the channel, and this is exactly the opposite of the assumption discussed for the calculation of the distribution of Γ_I. The ergodic rate calculation applies, for example, to a scenario where a given user is moving over a space-filling curve trajectory through the deployment region, traveling fast enough that, on the link between the user and any BS, we may assume independent block fading over successive intervals of duration T.

Unfortunately, when averaged over all PPP locations of the BSs, i.e. when averaged over all states of all channels of the system, the baseband equivalent complex total interference signal from all BSs is not Gaussian but symmetric alpha-stable (Gulati *et al.*, 2010). This has also been verified by simulation in a single-tier network (Aljuaid & Yanikomeroglu, 2010). Further, this distribution has characteristic exponent <2, hence it has no first or second moment. This implies that the achievable rate on the downlink to any user over a transmission duration that is much longer than T cannot even be computed analytically from the Shannon formula using the results in Lapidoth (1996). We do not consider the ergodic rate further in this book.

7.2.2 Spectral efficiency of an HCN

In Section 7.2.1, we were careful to talk of a link to an *arbitrarily located user*. This is because the expressions for the distribution of the SINR derived in Sections 5.3 and 6.4 are really for the SINR at an arbitrary *location*. In particular, these results on the distribution of the SINR do not depend upon the point pattern of the users in the system.

In order to talk of the spectral efficiency of a tier, however, it is more convenient to consider the problem from the perspective of a transmitter, namely an *arbitrary BS* in that tier. Since a BS transmits only to a specific subset of users (namely those users for which this BS is the serving BS), we begin by studying the spectral efficiency of the link from an arbitrary BS to a selected served user.

Spectral efficiency of the link from an arbitrary BS to a specific served user

Let us again consider a deployment of BSs modeled by a PPP Φ, and an arbitrary BS $b \in \Phi$. Denote the set of users whose serving BS is b by $\mathcal{U}_b$. Consider one specific served user, say $u_b \in \mathcal{U}_b$, and suppose the SINR at u_b when it receives from its serving BS b is $\Gamma(u_b)$. Note that the distribution of $\Gamma(u_b)$ is not the same as that of Γ in Section 7.2.1, because u_b is not chosen arbitrarily. In fact, $\Gamma(u_b)$ is the SINR at the specific user $u_b \in \mathcal{U}_b$, which itself is determined according to some serving BS selection criterion.

At each transmit interval, the BS b transmits to some user selected from the set $\mathcal{U}_b$ by the *scheduling policy*. These scheduling policies are usually chosen to maximize some *network utility function* (or some approximation thereof if the quantities needed to compute this network utility function are not available at the BSs or at a central controller that can then transmit the computed schedules to the BSs). For example, if the network utility is the sum of the instantaneous rates to the selected served users (where the sum is over all BSs in the network), this is achieved when each BS simply transmits to the served user with the highest instantaneous SINR. On the other hand, if the network utility is the sum of the logarithm of the long-term rates to all served users, it turns out that each BS should select the next user to transmit to based on the so-called *proportional-fair scheduler* (PFS). The topic of scheduling is a very rich one, and an entire book can easily be devoted to it. Further, there are several widely used schedulers, and the resulting analysis, even if tractable, would be different for each. An example of a (partial) analysis of the mean throughput under PFS is outlined in Problem 7.2.

Instead of considering the details of one or more schedulers, we consider a simple model where, at the instant at which we analyze the system, the BS b of interest chooses one user *at random* from the set of served users $\mathcal{U}_b$. This is mathematically equivalent to observing the operation of the so-called *round robin scheduler* (RRS) at the BS b at a randomly selected transmit interval. Hence we may also say that our analysis is for the RRS at each BS.

Spectral efficiency of a BS

With the above model, the *spectral efficiency of an arbitrary BS* in Φ is defined as the (unconditional) spectral efficiency of the link from this BS (say b) to a user U_b that is *randomly chosen* from the set of served users $\mathcal{U}_b$:

$$\mu_C(b) \equiv \mathbb{E}\{\log_2[1 + \Gamma(U_b)]\}, \quad b \in \Phi, \quad \mathbb{P}\{U_b = u_b\} = \frac{1}{|\mathcal{U}_b|}, \; u_b \in \mathcal{U}_b. \tag{7.6}$$

Remark 7.3 The spectral efficiency calculation requires the calculation of the SINR on the link from an arbitrary BS to a randomly selected served user. The restriction to a *served* user is important, because it enforces the *serving tier selection criterion*.

Remark 7.4 Note that with RRS, the BS b would transmit in turn to each $u_b \in \mathcal{U}_b$. Then, conditioned on the set of distances between BSs and the served users in $\mathcal{U}_b$, and the fade attenuations on all links between these BSs and served users, the total instantaneous rate of transmissions from b is $\sum_{u_b \in \mathcal{U}_b} \log_2[1 + \Gamma(u_b)]$ over $|\mathcal{U}_b|$ "uses" of the wireless "channel." Therefore, the spectral efficiency of transmissions from b, conditioned on these distances and fade attenuations, is given by

$$\frac{1}{|\mathcal{U}_b|} \sum_{u_b \in \mathcal{U}_b} \log_2[1 + \Gamma(u_b)] = \mathbb{E}\{\log_2[1 + \Gamma(U_b)]\},$$

where the expectation is over the distribution of the selected user U_b for transmission from b at each transmit interval, and this distribution is random and uniform over $\mathcal{U}_b$. Taking the expectation over the distances and fade attenuations then gives the unconditional spectral efficiency of transmissions from the BS b using the RRS to be (7.6).

Remark 7.5 Suppose that the candidate serving BSs in the tiers for each user are the ones that are nearest to that user. Then, we know that the distance from an *arbitrary*

BS to a randomly selected served user has a PDF that is well approximated by (6.6). Further, from the discussion in Section 6.4.5, we know that the similarity of (6.6) to (6.4) permits us to obtain the distribution of $\Gamma(u_b)$ in (7.6) (and thus the spectral efficiency $\mu_C(b)$) when an *arbitrary BS b transmits to a randomly selected served user* by using our previous results on the distribution (7.1) of the SINR at an *arbitrarily located user*, provided we change the distribution of the distance from this arbitrarily located user to its *serving BS* from (6.4) to (6.6).

In short, $\mu_C(b)$ for an arbitrary BS $b \in \Phi_i$, the ith tier of a multi-tier network, can be calculated by modifying (7.3) as follows:

$$\mu_C(b) = \frac{1}{\ln 2}\int_0^\infty \frac{1}{1+\gamma}\mathbb{P}\{\Gamma_i > \gamma, I = i\}\mathrm{d}\gamma, \quad b \in \Phi_i, \quad i = 1, \ldots, n_{\text{open}}, \tag{7.7}$$

where $\mathbb{P}\{\Gamma_I > \gamma\}$ in (7.3) is replaced by $\mathbb{P}\{\Gamma_I > \gamma, I = i\}$ in (7.7) because we actually want the spectral efficiency of an arbitrary BS in tier i, so, in the equivalent calculation for an arbitrarily located user in the network, we require this user to be served by (the candidate serving BS of) tier i.

The probability $\mathbb{P}\{\Gamma_i > \gamma, I = i\}$ is given by Theorems 5.10 and 5.11 for fixed transmit powers and (6.9) when the BSs use OLPC, with the PDF of the distance R_i to the serving BS (in tier i) given by (6.6). However, note that we are now performing the calculation of the SINR distribution at this arbitrarily located user, and the set of candidate serving BSs are the BSs from the open tiers that are nearest to this user location. Thus, in (7.7), the distributions of the distances from this arbitrarily located user to the *candidate serving BSs* of the *non-serving* tiers should continue to be taken as (6.4) for each tier i other than the serving tier.

Spectral efficiency of a tier and of the entire multi-tier HCN

Following the discussion in Section 7.2.1, we may view the spectral efficiency of an arbitrary BS (say b) as the *area averaged* mean downlink rate to a randomly selected user served by b, with the difference that the "area" is now the region served by b (e.g. a Voronoi cell with nucleus b if the candidate serving BS in each open tier is the one nearest to the user). We assume that each user in the system is served by at most one BS at each transmit interval. Then each BS in a tier can serve one user at each transmit interval. If the locations of the BSs in tier i are modeled by a homogeneous PPP Φ_i with density λ_i, $i = 1, \ldots, n_{\text{open}}$, then there are on average λ_i BSs belonging to tier i per unit area. It follows that the area-averaged mean downlink rate to randomly selected users served by (the BSs of) a *tier i* of the HCN is $\mu_{C,i} \equiv \lambda_i \mu_C(b)$, where b is an arbitrary BS of tier i, and $\mu_C(b)$ is given by (7.6). We refer to $\mu_{C,i}$ as the spectral efficiency of the tier i, $i = 1, \ldots, n_{\text{open}}$.

Since the BSs in the open tiers can serve users simultaneously in every transmit interval, it follows that the area-averaged mean downlink rate to randomly selected users served by all the BSs of the multi-tier HCN is $\mu_{C,\text{tot}} \equiv \sum_{i=1}^{n_{\text{open}}} \mu_{C,i}$. We will call $\mu_{C,\text{tot}}$ the spectral efficiency of the HCN.

7.2.3 Application: spectral efficiency of a macro-pico LTE HCN with eICIC

We turn again to the two-tier macro-pico LTE-like HCN described in Section 6.4.3. As stated there, we assume perfect time synchronization across the macro BSs. Further, we assume that the macros transmit no power at all during the ABSs. We also neglect thermal noise and assume i.i.d. Rayleigh fading on all links from any BS in any tier to any user.

A UE in the heterogeneous two-tier network comprising the picocell overlay Φ_2 on the underlying macrocellular network Φ_1 is served by either the nearest pico BS or the nearest macro BS. We know that we can compute the desired spectral efficiencies for each tier by reusing our earlier results on the distribution of the SINR at an arbitrarily located UE in the HCN, with the change that the PDF of the distances of this UE from the actual serving BS is changed from (6.4) to (6.6).

Range expansion and ABSs

Consider an arbitrarily located UE in a two-tier macro-pico HCN. Suppose the distance of the UE from the candidate serving (i.e. nearest) macro BS is R_1, and that from the candidate serving (i.e. nearest) pico BS is R_2. A *range expansion bias* (REB) θ is applied to the mean received pilot powers at the UE from the candidate serving macro and pico BSs in order to favor selection of the picocell tier to serve the UE as follows: the *serving* BS for the UE is the candidate serving pico BS if

$$\theta \frac{K_2 P_2^{\text{pilot,tx}}}{R_2^\alpha} > \frac{K_1 P_1^{\text{pilot,tx}}}{R_1^\alpha} \Leftrightarrow \tau R_2 > R_1,$$

and it is the candidate serving macro BS otherwise, where θ is the bias and

$$\tau = \sqrt{\frac{\lambda_1}{\lambda_2}\theta^{2/\alpha}\beta}, \quad \beta = \frac{\lambda_2}{\lambda_1}\left(\frac{K_2}{K_1}\frac{P_2^{\text{pilot,tx}}}{P_1^{\text{pilot,tx}}}\right)^{2/\alpha}.$$

Recall that a UE that has been offloaded (by means of range expansion) onto a picocell, which is in fact not the strongest received BS, experiences strong interference from the macro BS nearest to the UE. To mitigate this interference and still retain the benefits of offloading from the macro network to the small cells, the eICIC scheme proposes that the macro BSs should not transmit any power (other than that required for essential control signaling) on certain subframes. These *almost-blank subframes* (ABSs) are interspersed with normal subframes in a regular pattern. In our model, we make the idealized assumption that the macro BSs transmit no power at all during ABSs. Let the fraction of ABSs be η_{ABS}. Then the instantaneous transmit power of any macro BS is a random variable given by either zero, if in an ABS, or $P_{1,\text{reg}}^{\text{tx}}$ if not:

$$P_1^{\text{tx}} = (1 - \epsilon_{\text{ABS}}) \times P_{1,\text{reg}}^{\text{tx}} + \epsilon_{\text{ABS}} \times 0 = (1 - \epsilon_{\text{ABS}}) P_{1,\text{reg}}^{\text{tx}}, \tag{7.8}$$

where $\epsilon_{\text{ABS}} \sim \text{Bin}(1, \eta_{\text{ABS}})$. The assumption that macro BSs transmit no power at all during ABSs makes the macro BS tier effectively absent during ABSs, which turns the network into a single-tier one during ABSs. This means that we cannot use Theorem 6.1 and apply the fixed-power SINR distribution results of Section 5.3.1 with $P_1^{2/\alpha}$ replaced by $K_1^{2/\alpha}\, \mathbb{E}[(P_1^{\text{tx}})^{2/\alpha}]$ because the results in Section 5.3.1 assume the network tiers remain unchanged throughout.

For simplicity, we also assume that the pilot channel powers are related to the traffic channel powers in the same way in both tiers:

$$\frac{P_1^{\text{pilot,tx}}}{P_{1,\text{reg}}^{\text{tx}}} = \frac{P_2^{\text{pilot,tx}}}{P_2^{\text{tx}}},$$

so that we have

$$\beta = \frac{\lambda_2}{\lambda_1}\left(\frac{K_2}{K_1}\frac{P_2^{\text{tx}}}{P_{1,\text{reg}}^{\text{tx}}}\right)^{2/\alpha} = \frac{\lambda_2 P_2^{2/\alpha}}{\lambda_1 P_1^{2/\alpha}}, \quad \theta = \tau^{\alpha}\frac{P_1}{P_2}, \quad P_1 \equiv K_1 P_{1,\text{reg}}^{\text{tx}},\ P_2 \equiv K_2 P_2^{\text{tx}}.$$

Spectral efficiency of the macro tier

From (7.7) and the criterion for serving BS selection, the spectral efficiency of the macro tier is given by

$$\begin{aligned}\mu_{C,1} &= \frac{\lambda_1}{\ln 2}\int_0^{\infty}\frac{\mathbb{P}\{\Gamma_1 > \gamma, R_2 > \tau R_1\}}{1+\gamma} \\ &= \frac{\lambda_1}{\ln 2}\int_0^{\infty}\frac{\mathrm{d}\gamma}{1+\gamma}\int_0^{\infty}\mathrm{d}r f_{R_1}(r)\int_{\tau r}^{\infty}\mathrm{d}r' f_{R_2}(r')\,\mathbb{P}\{\Gamma_1 > \gamma \mid R_1 = r, R_2 = r'\},\end{aligned} \tag{7.9}$$

where we note from our earlier discussion that, as the macro tier is the serving tier, we should use the following PDF of R_1:

$$f_{R_1}(r) = 2\pi\lambda_1 c r \exp\left(-\pi\lambda_1 c r^2\right), \quad r \geq 0, \quad c = 1.25,$$

while the PDF of R_2, the distance to the candidate serving BS of the non-serving tier, remains as follows:

$$f_{R_2}(r') = 2\pi\lambda_2 r' \exp\left[-\pi\lambda_2 (r')^2\right], \quad r' \geq 0.$$

Further, from (7.8) we have

$$\begin{aligned}\mathbb{P}\{\Gamma_1 > \gamma \mid R_1 = r, R_2 = r'\} &= (1-\eta_{\text{ABS}})\,\mathbb{P}\{\Gamma_1^{\text{not ABS}} > \gamma \mid R_1 = r, R_2 = r'\} \\ &= (1-\eta_{\text{ABS}})\exp\left\{-\pi\lambda_1 r^2\gamma^{2/\alpha}\left[G_\alpha\left(\frac{1}{\gamma^{2/\alpha}}\right) + \beta G_\alpha\left(\frac{\pi\lambda_2 (r')^2}{\beta\gamma^{2/\alpha}\pi\lambda_1 r^2}\right)\right]\right\},\end{aligned} \tag{7.10}$$

where we have used (5.7) with $N_0 = 0$ and $k = 1$, and $G_\alpha(\cdot)$ is given by (4.29). To obtain the spectral efficiency of the macro tier, we can then substitute (7.10) into (7.9).

Spectral efficiency of the pico tier

Similarly, the spectral efficiency of the pico tier is given by

$$\mu_{C,2} = \frac{\lambda_2}{\ln 2}\int_0^{\infty}\frac{\mathrm{d}\gamma}{1+\gamma}\int_0^{\infty}\mathrm{d}r' f_{R_2}(r')\int_{r'/\tau}^{\infty}\mathrm{d}r f_{R_1}(r)\,\mathbb{P}\{\Gamma_2 > \gamma \mid R_1 = r, R_2 = r'\}, \tag{7.11}$$

where this time the pico tier is the serving tier, so the PDF of R_2 should be taken as

$$f_{R_2}(r') = 2\pi\lambda_2 c r' \exp\left[-\pi\lambda_2 c (r')^2\right], \quad r' \geq 0, \quad c = 1.25,$$

while the PDF of R_1, the distance to the candidate serving BS of the non-serving tier, is given by

$$f_{R_1}(r) = 2\pi\lambda_1 r \exp\left(-\pi\lambda_1 r^2\right), \quad r \geq 0.$$

Now,

$$\begin{aligned}
&\mathbb{P}\{\Gamma_2 > \gamma \mid R_1 = r, R_2 = r'\} \\
&\quad = (1 - \eta_{\mathrm{ABS}})\, \mathbb{P}\{\Gamma_2^{\text{not ABS}} > \gamma \mid R_1 = r, R_2 = r'\} \\
&\qquad + \eta_{\mathrm{ABS}}\, \mathbb{P}\{\Gamma_2^{\mathrm{ABS}} > \gamma \mid R_1 = r, R_2 = r'\} \\
&\quad = (1 - \eta_{\mathrm{ABS}}) \exp\left\{-\pi\lambda_2 (r')^2 \gamma^{2/\alpha} \left[G_\alpha\left(\frac{1}{\gamma^{2/\alpha}}\right) + \frac{1}{\beta} G_\alpha\left(\frac{\pi\lambda_1 r^2}{\frac{1}{\beta}\gamma^{2/\alpha}\pi\lambda_2 (r')^2}\right)\right]\right\} \\
&\qquad + \eta_{\mathrm{ABS}} \exp\left[-\pi\lambda_2 (r')^2 \gamma^{2/\alpha} G_\alpha\left(\frac{1}{\gamma^{2/\alpha}}\right)\right],
\end{aligned} \tag{7.12}$$

where we have used (5.7) with $k = 1$, swapping the labels of tiers 1 and 2 for $\mathbb{P}\{\Gamma_2^{\text{not ABS}} > \gamma \mid R_1 = r, R_2 = r'\}$, and (4.28) for $\mathbb{P}\{\Gamma_2^{\mathrm{ABS}} > \gamma \mid R_1 = r, R_2 = r'\} = \mathbb{P}\{\Gamma_2^{\mathrm{ABS}} > \gamma \mid R_2 = r'\}$ for the single-tier (pico-only) network that exists during ABSs.

Once more, we can substitute (7.12) into (7.11) in order to obtain the spectral efficiency of the pico tier.

Finally, the spectral efficiency of the macro-pico HCN is given by $\mu_{C,\mathrm{tot}} = \mu_{C,1} + \mu_{C,2}$.

Numerical results

We note here that if we had, for example, an alternative single-tier homogeneous PPP deployment ($n_{\mathrm{open}} = n_{\mathrm{tier}} = 1$) with density λ, the same path-loss exponent, i.i.d. Rayleigh fading on all links, and the same rule that any UE be served by the nearest BS of this deployment, the spectral efficiency of such a deployment is given by (see Problem 7.3)

$$\mu_{C,\mathrm{alt}} = \lambda \frac{c}{\ln 2} \int_0^\infty \frac{\mathrm{d}\gamma}{(1+\gamma)\left[c + \gamma^{2/\alpha} G_\alpha\left(\gamma^{-2/\alpha}\right)\right]}, \tag{7.13}$$

which does not depend upon the (fixed) BS transmit powers or the intercept of the path-loss model.

We set $\alpha = 4$, $K_2/K_1 = -12$ dB, and $\eta_{\mathrm{ABS}} = 0.25$. We set the density of the baseline macro tier (tier 1) to unity ($\lambda_1 = 1$) and plot $\mu_{C,1}$, $\mu_{C,2}$, and $\mu_{C,\mathrm{tot}}$ vs. $P_2^{\mathrm{tx}}/P_{1,\mathrm{reg}}^{\mathrm{tx}}$ for various choices of θ in Figure 7.1 with $\lambda_2 = 10$. For comparison, we also plot the spectral efficiency (7.13) for two alternative single-tier deployments: one a standalone macro-only tier, with the same density $\lambda_1 = 1$ and transmit power $P_{1,\mathrm{reg}}^{\mathrm{tx}}$ as the macro tier of the HCN (but no ABS), the other a standalone small-cell tier with the same

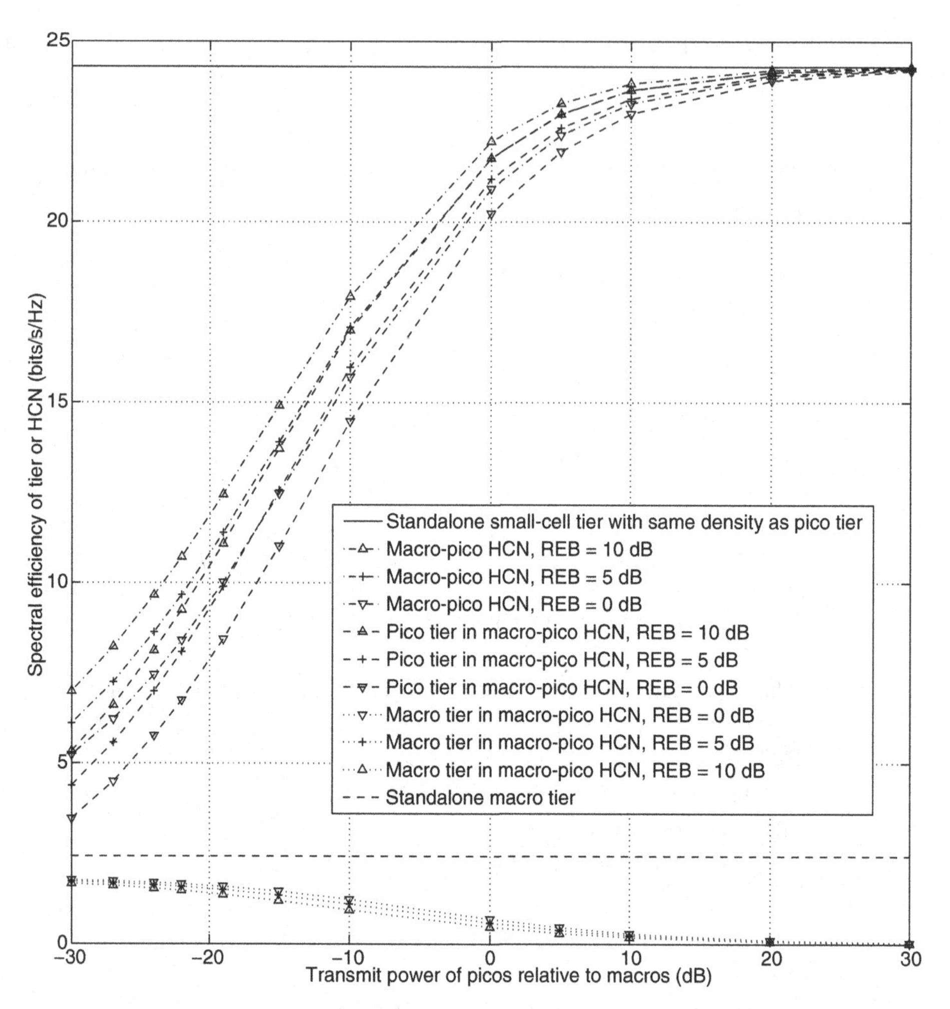

Figure 7.1 Spectral efficiency of a standalone single-tier (macro-only) deployment, the total spectral efficiency of a two-tier macro-pico HCN, and the spectral efficiency of a standalone small-cell single-tier deployment, vs. the relative transmit powers $P_2^{\rm tx}/P_{1,\rm reg}^{\rm tx}$ when the small-cell and pico-tier densities are ten times that of the (standalone or HCN) macro tier, and the macro tier in the two-tier HCN transmits no power during ABSs of fractional duration $\eta_{\rm ABS} = 0.25$. Three plots are given for each deployment, corresponding to the range expansion bias (REB) θ taking the values 0 dB, 5 dB, and 10 dB. For comparison, the spectral efficiency of the macro and pico tiers of the two-tier HCN are plotted separately. Thermal noise is neglected, and the path-loss exponent is $\alpha = 4$.

transmit power P_2 and density $\lambda_2 = 10$ as the pico tier of the HCN. From the spectral efficiency curves, the benefits of spatial reuse of resources in a dense small-cell tier are quantified. In particular, we see that, in the macro-pico HCN, the pico tier's spectral efficiency approaches that of the standalone small-cell tier, with the same density, only when the pico BSs are transmitting with 20 dB or greater power than the macro BSs, and therefore carry nearly all the traffic in the network. This is also seen from the fact that the spectral efficiency of the macro tier in the macro-pico HCN drops to nearly zero as the pico BSs' transmit power increases.

The poorer spectral efficiency of the macro-pico HCN occurs because of interference between the tiers and (in the case of the macro tier) the zero-power ABS intervals. The spectral efficiency advantage of the standalone small-cell tier makes it a viable candidate for a greenfield deployment, say on a dedicated band that is not shared with the macro tier (Mukherjee & Ishii, 2013).

Problems

7.1 Prove (7.4) for any non-negative continuous-valued random variable X with PDF $f_X(\cdot)$. *Hint*:

$$\mathbb{E}[X] = \int_0^\infty t f_X(t)\mathrm{d}t = \int_0^\infty \int_0^t \mathrm{d}x\, f_X(t)\mathrm{d}t = \int_0^\infty \int_x^\infty f_X(t)\mathrm{d}t\,\mathrm{d}x.$$

7.2 In this problem we show an interesting application of the canonical probability for the calculation of mean throughput to a served user under proportional-fair scheduling. Consider a single-tier network where each user is served by the nearest BS. Consider a BS b serving a set of N users $\mathcal{U}_b = \{u_{b,1}, \ldots, u_{b,N}\}$. Let Γ_i denote the instantaneous SINR at user $u_{b,i}$ when receiving from BS b, $i = 1, \ldots, N$, and let $C_i \equiv \log_2(1 + \Gamma_i)$ be the corresponding instantaneous rate at $u_{b,i}$. Suppose the distances of the users in $\mathcal{U}_b$ from BS b are denoted $R_{b,1}, \ldots, R_{b,N}$, respectively. Assume the users are stationary, so these distances are assumed to be fixed at $R_{b,i} = r_{b,i}$, $i = 1, \ldots, N$, and all probabilities and expectations are calculated conditioned on the event $\boldsymbol{R}_b = \boldsymbol{r}_b$, where $\boldsymbol{R}_b = [R_{b,1}, \ldots, R_{b,N}]^\top$ and $\boldsymbol{r}_b = [r_{b,1}, \ldots, r_{b,N}]^\top$. We also assume i.i.d. fading on all links from b to the users in $\mathcal{U}_b$, so $C_1, \ldots, C_N$ are independent.

Assume that the system has reached its long-term steady state. At each transmit interval, the proportional-fair scheduler (PFS) selects a user $u_{b,I} \in \mathcal{U}_b$ to be transmitted to by the BS b. Under the steady-state assumption, we may write (Liu, Zhang & Leung, 2011, eqn. (12))

$$I = \arg\max_{1 \le j \le N} \frac{C_j}{\mathbb{E}[C_j\, 1_{\{j\}}(I) \mid R_{b,1} = r_{b,1}, \ldots, R_{b,N} = r_{b,N}]}.$$

(1) Define $\mu_j \equiv \mathbb{E}[C_j\, 1_{\{j\}}(I) \mid \boldsymbol{R}_b = \boldsymbol{r}_b]$, $j = 1, \ldots, N$. Observe that μ_j is the mean throughput to the jth served user (conditioned on the locations of these users relative to the serving BS). Prove that

$$\mu_j = \sum_{n=1}^{N} (-1)^{n-1} \sum_{\substack{j_1, \ldots, j_n:\\ j_1 \equiv j,\\ j_1 \neq j_l, l=2,\ldots,N,\\ 1 \le j_2 < \cdots < j_N}} \int_0^\infty \mathbb{P}\Big\{C_j > x, C_{j_l} > \frac{\mu_{j_l}}{\mu_j} C_j,\ l = 2, \ldots, n \;\Big|\; \boldsymbol{R}_b = \boldsymbol{r}_b\Big\}\mathrm{d}x. \tag{7.14}$$

Hint: $C_j 1_{\{j\}}(I)$ is the rate from b to $u_{b,j}$ when the latter is selected by the PFS, so we can write (Liu *et al.*, 2011, eqn. (16))

$$\begin{aligned}\mu_j &= \int_0^\infty t f_{C_j \mid \boldsymbol{R}_b}(t \mid \boldsymbol{r}_b) \prod_{\substack{i=1\\ i\neq j}}^{N} \mathbb{P}\left\{C_i \le \frac{\mu_i}{\mu_j} t \mid \boldsymbol{R}_b = \boldsymbol{r}_b\right\} \mathrm{d}t \\ &= \int_0^\infty \int_0^t f_{C_j \mid \boldsymbol{R}_b}(t \mid \boldsymbol{r}_b) \prod_{\substack{i=1\\ i\neq j}}^{N} \mathbb{P}\left\{C_i \le \frac{\mu_i}{\mu_j} t \mid \boldsymbol{R}_b = \boldsymbol{r}_b\right\} \mathrm{d}x\, \mathrm{d}t \\ &= \int_0^\infty \int_x^\infty f_{C_j \mid \boldsymbol{R}_b}(t \mid \boldsymbol{r}_b) \prod_{\substack{i=1\\ i\neq j}}^{N} \mathbb{P}\left\{C_i \le \frac{\mu_i}{\mu_j} t \mid \boldsymbol{R}_b = \boldsymbol{r}_b\right\} \mathrm{d}t\, \mathrm{d}x \\ &= \int_0^\infty \mathbb{P}\left\{C_j > x, C_i \le \frac{\mu_i}{\mu_j} C_j,\ i \in \{1, \ldots, N\} \setminus \{j\} \mid \boldsymbol{R}_b = \boldsymbol{r}_b\right\} \mathrm{d}x,\end{aligned}$$

and then use the inclusion-exclusion formula to expand the probability in the integral, following the same steps as in (5.73) and (5.75), and using the summation notation in Remark 5.10.

(2) Follow similar steps to the derivation of (5.76) to show that, for each $n \ge 2$ and each $j_1, \ldots, j_n$, conditioned on $\boldsymbol{R}_b = \boldsymbol{r}_b$,

$$C_j > x, C_{j_l} > \frac{\mu_{j_l}}{\mu_j} C_j,\ l = 2, \ldots, n \quad \Leftrightarrow \quad \mathbf{A}_n(\boldsymbol{\theta})\boldsymbol{X} > x\boldsymbol{e}_1^{(n)},$$

where $\boldsymbol{X} \equiv [C_j, C_{j_2}, \ldots, C_{j_n}]^\top$ and $\mathbf{A}_n(\boldsymbol{\theta})$ is a Z-matrix given by

$$\mathbf{A}_n(\boldsymbol{\theta}) = \mathbf{I}_n + \left[\boldsymbol{e}_1^{(n)} - \boldsymbol{\theta}\right]\left(\boldsymbol{e}_1^{(n)}\right)^\top, \quad \boldsymbol{\theta} \equiv [\theta_1, \ldots, \theta_n]^\top = \left[1, \frac{\mu_{j_2}}{\mu_j}, \ldots, \frac{\mu_{j_n}}{\mu_j}\right]^\top.$$

(3) Thus, in theory it is possible to calculate each of the probabilities in (7.14) using Lemma 2.3 provided the PDF of each C_i (conditioned on $R_{b,i} = r_{b,i}$, and obtained from the corresponding conditional distribution of Γ_i given by Theorem 4.9) can be written in the form (2.17). Then (7.14) yields N fixed-point equations among $\mu_1, \ldots, \mu_N$, which would need to be solved numerically to obtain the throughputs to these served users (conditioned on their distances from the serving BS).

7.3 Prove that, for a single-tier homogeneous PPP Φ with density λ, the spectral efficiency of the tier when the serving BS for each user is the nearest one is given by (7.13) when $N_0 = 0$. *Hint*: If R is the distance of the nearest BS to an arbitrarily located user, the spectral efficiency of the tier is given by

$$\begin{aligned}&\lambda \int_0^\infty \mathbb{E}\left[\log_2(1+\Gamma) \mid R = r\right] f_R(r) \mathrm{d}r \\ &= \frac{\lambda}{\ln 2} \int_0^\infty \frac{\mathrm{d}\gamma}{1+\gamma} \int_0^\infty \mathrm{d}r\, \mathbb{P}\{\Gamma > \gamma \mid R = r\} f_R(r),\end{aligned}$$

where $\mathbb{P}\{\Gamma > \gamma \mid R = r\}$ is given by (4.28), and $f_R(r) = 2\pi\lambda c r \exp(-\pi\lambda c r^2)$, $r \ge 0$.

7.3 Energy efficiency

The expansion of cellular networks has given rise to concerns, both from network operators interested in reducing energy costs, and from all parties interested in reducing the carbon footprints of such networks (Fehske *et al.*, 2011). The key metric here is a measure of the spectral efficiency of the network per unit of power (either total consumed power, or power transmitted over the air), called *energy efficiency*. There are several different definitions of energy efficiency, all variations of this basic concept (Abdulkafi *et al.*, 2012). However, the most widely used definition is the ratio of (area-averaged) spectral efficiency of the network to the transmit power consumption per unit area (Bossmin, 2012).

In the case where the tiers of the HCN have BSs whose locations may be modeled by homogeneous PPPs, the energy efficiency of each tier is simply the ratio of spectral efficiency of an arbitrary BS in that tier to its power consumption. We note here that the consumed power is typically significantly greater than the power transmitted over the air, and it is the total consumed power that is of greater importance in determining the energy efficiency. The calculation of total power consumption requires a model that takes into account the hardware design of the BSs and the requirements of the standards. At the present time, the consensus in the industry is to use the following model (Auer *et al.*, 2011a) for the total power consumption P_{tot} of a BS:

$$P_{\text{tot}} = \begin{cases} \Delta_p P_{\text{max}}^{\text{tx}} + P_0, & \text{if the BS is active,} \\ P_{\text{sleep}}, & \text{if the BS is in sleep mode,} \end{cases}$$

where P_0 is the power consumed by the BS when it transmits the least possible power over the air, $P_{\text{max}}^{\text{tx}}$ is the maximum possible (over-the-air) transmit power of the BS, and $0 < \Delta_p < 1$ is the slope of the load-dependent power consumption, to model the fact that the BS's transmit power scales up with load until it reaches its maximum value.

A framework to compute the power consumption per unit area of a network based on modeling of the power consumption, user density, user traffic demand (load), and BS density for a single-tier network is provided in Auer *et al.* (2011b) and employed to compute energy efficiency in Suryaprakash *et al.* (2012). All that remains to compute the energy efficiency is to compute the spectral efficiency of the network, and the results in this chapter show us how to do so. We do not explore the topic of power consumption modeling further in this book.

8 Closing thoughts: future heterogeneous networks

8.1 Introduction

We have already discussed why the wireless cellular networks of the future are going to be *heterogeneous* in nature, with a mix of BSs with different capabilities and characteristics. The spectral efficiency advantages obtained by dense spatial reuse of resources are driving the deployment of small-cell (i.e. small in terms of range or cell size relative to macrocells) tiers as overlays on the existing macrocellular tier.

It is interesting to observe that, as cell sizes shrink, they become comparable to the range of WiFi (IEEE 802.11) *access points*. This has led to a resurgence of interest in the *peer-to-peer* (P2P) mode of operation of WiFi, now enhanced and extended to cellular links as *device-to-device* (D2D) communication, as a means of further enhancing area-wide spectral efficiency. Thus, future heterogeneous wireless networks are also expected to support such D2D communication. As a result, the wireless cellular networks of the future will be not only heterogeneous, but also not even fully *cellular* – direct transmissions between user devices will be not only permitted, but also enabled by the BSs in order to support ever greater numbers of simultaneously active links within each cell.

8.2 Analysis of a network with D2D links

With the renewal of interest in D2D links, it can be said that the analysis of wireless networks has now come full circle, back to its original roots in P2P links (Hunter *et al.*, 2008; Weber & Andrews, 2012). However, a wireless cellular network permitting D2D transmissions among user devices exhibits some novel features not seen before.

(1) The D2D links will use *licensed* spectrum. In conventional cellular networks, this spectrum has a sole licensee, the network operator, who has often paid a high price at a spectrum auction for exclusive use of this spectrum. All transmissions on this spectrum in the region over which this operator holds the license must take place subject to the approval of the operator (i.e. by BSs belonging to the operator, and by user devices belonging to subscribers of this operator's service).

However, when user devices (belonging to subscribers of the licensed operator's service) engage in direct transmissions to one another, possibly not subject to the control of the operator's network infrastructure (i.e. the BSs), the operator has lost an important element of control on the usage of its expensively acquired licensed spectrum. Since it is inconceivable that the D2D transmissions be permitted to disrupt the operation of the cellular links, the former class of transmissions must necessarily use only those resources that are not in use by cellular transmissions (i.e. between BSs and user devices). This may be enabled by coordinating all D2D transmissions at the BSs. On the other hand, a much simpler network architecture results if the user devices are permitted to coordinate such D2D transmissions autonomously so as not to interfere with ongoing cellular transmissions. This leads us directly to topics on *cognitive radio*, with the D2D transmitters treated as *secondary users* while the transmitters and receivers of the conventional cellular links are treated as the *primary users*.

(2) D2D transmissions will have to employ time division duplexing (TDD), in contrast to the overwhelming majority of conventional cellular networks today, which employ frequency division duplexing (FDD). Though FDD on D2D links is not theoretically impossible, it is practically infeasible to obtain a pair of frequency bands via cognitive means, compared to finding an interval of time when no cellular transmissions are ongoing. The presence of TDD links in a cellular network can lead to unusual sources of interference, e.g. the interference at a (cellular) user now has a component from other (D2D) *users*, in addition to the interference from non-serving BSs. The effects of time synchronization among the BSs and users of the system will need to be investigated carefully, as lack of time synchronization could have serious effects on the performance of TDD links.

(3) In the future, the D2D transmitters and receivers may not belong to human subscribers of the operator's service. Note that, in the present cellular network, most of the traffic in the system is human initiated, e.g. making a call, browsing a website, downloading/uploading a video, etc. However, there is a component of it that is automatic, e.g. "push" messaging and email services and "app" updates. These are simple examples of what is projected to become the majority of traffic volume over wireless networks in the next decade, namely *machine-type communications* (MTCs). It is expected that in the wireless networks of the future, the majority of the traffic will be machine initiated and directed to other machines, without any human involvement or intervention, just as is true today with wireline networks. Such MTC cases include not only the well-studied examples of sensor networks for telemetry, vending-machine stocking, inventory management, etc., but also new scenarios involving device discovery, application discovery, caching, storage, and forwarding, all incorporating as-yet-unknown economic models for pricing of such P2P services.

In short, the presence of D2D links within conventional cellular networks will tie the field of cellular wireless networks to that of ad-hoc, sensor, and cognitive radio networks. The study of D2D transmissions within cellular networks is in its infancy, but a good idea of the challenges, analytic techniques, and results may be obtained from Lee *et al.,*[1] Lin, Andrews & Ghosh,[2] and Caire, Adhikary & Ntranos (2013).

8.3 The role of WiFi in future HCNs

Up to now, we have said nothing about the role of the IEEE 802.11 standard (popularly known as WiFi) in cellular wireless networks. This is because this role has not amounted to much so far, being limited to cellular network operators' attempts to offload their subscribers from the (macro-)cellular network onto the unlicensed spectrum of WiFi, offered through access points placed in so-called "hotspots" of high traffic demand. Unfortunately, this offload is not entirely seamless, in the sense that the subscriber cannot be handed off to such a WiFi network if encountering it for the first time. Access to such WiFi networks for the first time usually requires some sort of authentication, usually via a password (given offline to the subscriber by the cellular network operator provisioning or at least subsidizing the WiFi hotspot). In addition to this potential source of session interruption, there is also the fact that WiFi links are best-effort only and offer no guarantees of service. Cellular network operators seem to view WiFi as a "bonus" that is thrown in for free or nominal charge with the "premium" cellular service. The fact is, however, that the IEEE 802.11 standard, though hampered by its simple medium access control design, nonetheless has physical layer capabilities that match (or, in the case of the latest IEEE 802.11ac standard, exceed) that in the LTE standard (albeit over a smaller range than LTE). Cellular network operators have moved from dismissal of WiFi through grudging acceptance to at least partial acknowledgement of its role in future wireless networks, and are now working with the IEEE and the WiFi Alliance (the standards body and industry forum responsible for the 802.11 standard, respectively) to enable seamless handoff and offload from cellular to WiFi networks.

In short, WiFi will play an increasingly visible role in future *cellular* wireless deployments. In addition to offloading data from the cellular tiers, the WiFi infrastructure can also support D2D links over *unlicensed* spectrum, thereby obviating the complicated spectrum-allocation schemes that would otherwise be required to support D2D links over licensed spectrum shared with cellular links. However, the multiple-access method used in WiFi, called carrier-sensed multiple access, or CSMA, is very different from the orthogonal frequency division multiple access, or OFDMA (downlink), scheme used in cellular networks. The study of total interference at the receiver of a link in a dense WiFi network depends upon some additional parameters (compared to the analysis of a cellular network), such as the activity levels of the other users and access points, with the situation becoming more complicated if WiFi D2D transmissions are allowed. The

[1] N. Lee, X. Lin, J. G. Andrews & R. Health Jr., "Power control for D2D underlaid cellular networks: modeling, algorithm and analysis," *IEEE Journal on Selected Areas of Communications* (submitted for publication 2013). Available at http://arxiv.org/abs/1305.6161; last accessed May 27, 2013.

[2] X. Lin, J. G. Andrews & A. Ghosh, "A comprehensive framework for device-to-device communications in cellular networks," *IEEE Journal on Selected Areas of Communications* (submitted for publication 2013). Available at http://arxiv.org/abs/1305.4219; last accessed June 22, 2013.

WiFi tier may prove to be a better choice for offload of certain classes of traffic from the cellular network compared to other classes of traffic, and much work remains to be done on joint design of WiFi and cellular networks.

8.4 Evolution of the network infrastructure

The preceding discussion has touched upon only one aspect of a wireless communication system, namely the air-interface. However, the heterogeneous wireless networks of the future will also feature quite different network elements and infrastructure compared to the cellular networks of today. One technology that is an area of active interest by network operators is that of centralized baseband processing for all users' sessions within a wide area, with the appropriate signals to be transmitted to these served users being sent in baseband form to network elements that are now little more than antennas with power amplifiers attached to them. Such network elements are called remote radio heads (RRHs), and are usually assumed to be linked to the central entity by fast connections such as fiber optic lines. A variation of this idea is that of employing *relays* with no wired backhaul to the core network to enhance coverage in certain regions of the deployment area. Though many fundamental information-theoretic questions about relay networks remain unanswered, this area has experienced a burst of activity recently, though more from academic researchers than from network operators.

At any rate, the prospect of *coordinated transmission* to a user from multiple transmitters (or antennas) in the network is one that has excited the interest of many researchers, and RRHs and *distributed antenna systems* are a way to achieve this. In this case, a single user is "served" by multiple transmitters that may not be co-located, which considerably complicates the problem formulation for the analysis of the SINR distribution at this user.

8.5 New directions in analysis

This book has described analytic techniques to study topics such as coverage (SINR distribution) and capacity (spectral efficiency) in heterogeneous *cellular* networks. As the discussion makes clear, future wireless networks will not only be heterogeneous, but will also incorporate non-cellular links such as D2D transmissions, WiFi links, and relay transmissions. The next phase of our design of such systems for superior performance will require an understanding of the same issues of SINR and spectral efficiency (possibly combined with energy efficiency) in such networks of the future. Thus the methods and results described in this book are only a beginning. It will be interesting and exciting to see their application and extension to future heterogeneous wireless networks, and it is hoped that this book has both excited the reader's interest in these topics, and given the reader a foundation of tools and techniques on which to build in order to tackle such problems.

Appendix A Some common probability distributions

In this appendix, we define the probability distributions used in the book. We do not provide any detailed properties of these distributions here. Such details may be obtained from any of several standard references, e.g. Papoulis (1991).

A.1 Discrete distributions

A.1.1 Uniform distribution

This is what we mean when we talk of "randomly" selecting one object from a set of size n. If we label the objects in this set from 1 to n and let X denote the random variable representing the label of the selected object, the selection of an object "at random" corresponds to choosing a label with the following distribution:

$$X \sim \text{Unif}\{1, \ldots, n\} \Leftrightarrow \mathbb{P}\{X = k\} = \frac{1}{n}, \quad k = 1, \ldots, n.$$

A.1.2 Bernoulli distribution

A random variable X that takes only one of the values 0 or 1 is said to have a Bernoulli distribution with parameter p:

$$\mathbb{P}\{X = 1\} = p = 1 - \mathbb{P}\{X = 0\}.$$

The Bernoulli distribution can be used to model the outcome of a single experiment (called a "trial" in probability parlance), with 1 denoting success and 0 failure, say. For example, it can model the outcome of a single toss of a biased coin, with 1 denoting heads, say, and 0, tails.

A.1.3 Binomial distribution

The sum of n independent identically distributed (i.i.d.) Bernoulli random variables $X_1, \ldots, X_n$ (each with the same probability p of taking the value 1) is said to have the binomial distribution with parameters n and p:

$$X \sim \text{Bin}(n, p) \Leftrightarrow \mathbb{P}\{X = k\} = \binom{n}{k} p^k (1 - p)^{n-k}, \quad k = 0, 1, \ldots, n.$$

It follows that the Bernoulli distribution with parameter p is equivalent to the binomial distribution with parameters 1 and p, $\mathrm{Bin}(1,p)$. Thus $\mathrm{Bin}(n,p)$ is the distribution of the number of successes in n independent Bernoulli "trials." For the coin-tossing example, the $\mathrm{Bin}(n,p)$ distribution models the number of heads in n independent coin tosses.

A.1.4 Poisson distribution

The Poisson distribution with parameter λ, denoted $\mathrm{Poiss}(\lambda)$, is the $\mathrm{Bin}(n,\lambda/n)$ distribution where $p = \lambda/n$ for some fixed positive λ in the limit as $n \to \infty$:

$$X \sim \mathrm{Poiss}(\lambda) \Leftrightarrow \mathbb{P}\{X = k\} = \mathrm{e}^{-\lambda}\frac{\lambda^k}{k!}, \quad k = 0, 1, 2, \ldots.$$

A.1.5 Negative binomial distribution

The negative binomial distribution with parameters r and p, denoted $\mathrm{NB}(r,p)$, is the distribution of the number of successes in a sequence of independent Bernoulli trials that is conducted as long as required to obtain exactly r failures (where the probability of failure on each trial is $1-p$):

$$X \sim \mathrm{NB}(r,p) \Leftrightarrow \mathbb{P}\{X = k\} = \binom{r+k-1}{k}(1-p)^r p^k, \quad k = 0, 1, \ldots.$$

A.1.6 Generalized negative binomial distribution

The generalized negative binomial distribution, or Pólya distribution, with parameters $r > 0$ and p is just the negative binomial distribution generalized to the case where r is not an integer:

$$\mathbb{P}\{X = k\} = \frac{\Gamma(r+k)}{\Gamma(r)k!}(1-p)^r p^k, \quad k = 0, 1, \ldots.$$

A.2 Continuous distributions

A.2.1 Uniform distribution

The continuous uniform distribution is the counterpart of the continuous discrete distribution. Specifically, the continuous distribution on the interval $[a,b]$ is one where the PDF is constant over this interval and zero everywhere else:

$$X \sim \mathrm{Unif}[a,b] \Leftrightarrow f_X(x) = \begin{cases} 1/(b-a), & a \le x \le b, \\ 0, & \text{elsewhere.} \end{cases}$$

A.2.2 Normal or Gaussian distribution

A normal or Gaussian distribution with mean μ and variance σ^2 is one with the following PDF:

$$X \sim \mathcal{N}(\mu, \sigma^2) \Leftrightarrow f_X(x) = \frac{1}{\sqrt{2\pi\sigma^2}} \exp\left[-\frac{(x-\mu)^2}{2\sigma^2}\right], \quad -\infty < x < \infty.$$

The $\mathcal{N}(0, 1)$ distribution is called the *standard* or *unit* normal or Gaussian distribution.

A.2.3 Circularly symmetric complex Gaussian distribution

If X and Y are i.i.d. $\mathcal{N}(0, \sigma^2)$ for some $\sigma > 0$, then $Z = X + \mathrm{j}Y$ is said to have a *circularly symmetric complex Gaussian* distribution, written $\mathcal{CN}(0, 2\sigma^2)$. The $\mathcal{CN}(0, 1)$ distribution is called the *unit* circularly symmetric complex Gaussian distribution.

A.2.4 Rayleigh distribution

The Rayleigh distribution with *scale parameter* $\sigma > 0$ is the distribution of the magnitude $|Z|$ of a circularly symmetric complex Gaussian random variable $Z \sim \mathcal{CN}(0, 2\sigma^2)$. In other words, the distribution of $R \equiv \sqrt{X^2 + Y^2}$, where X and Y are i.i.d. $\mathcal{N}(0, \sigma^2)$ is Rayleigh with parameter σ and has the PDF

$$R \sim \text{Rayleigh}(\sigma) \Leftrightarrow f_R(r) = \frac{r}{\sigma^2} \exp\left(-\frac{r^2}{2\sigma^2}\right), \quad r \geq 0.$$

When we talk of *Rayleigh fading* on a wireless link, we mean that the in-phase and quadrature components of the *received signal* are i.i.d. zero-mean Gaussian random variables, so that the magnitude of the equivalent complex baseband signal is Rayleigh distributed. The Rayleigh(1) distribution is called the *unit* Rayleigh distribution.

A.2.5 Exponential distribution

The exponential distribution with mean 1, called the *unit* exponential distribution and denoted Exp(1), is the distribution of the random variable $U = R^2/2$, where $R \sim$ Rayleigh(1), i.e. $U = (X^2 + Y^2)/2$, where X and Y are i.i.d. $\mathcal{N}(0, 1)$:

$$U \sim \text{Exp}(1) \Leftrightarrow f_U(u) = \mathrm{e}^{-u}, \quad u \geq 0.$$

Thus, the unit exponential distribution is the distribution of the *normalized power* of a received complex baseband *signal* under unit Rayleigh fading. In other words, when the fading is Rayleigh, the *magnitude of the signal* is Rayleigh, and the *power* is exponential.

The exponential distribution with mean $\mu > 0$, denoted Exp(μ), is just the distribution of μU, where $U \sim$ Exp(1):

$$V \sim \text{Exp}(\mu) \Leftrightarrow f_V(v) = \frac{1}{\mu} \exp\left(-\frac{v}{\mu}\right), \quad v \geq 0.$$

A.2.6 Erlang distribution

The Erlang distribution with *shape* parameter $k \in \{1, 2, \dots\}$ and *scale* parameter $\mu > 0$ (or equivalently *rate* parameter $\lambda = 1/\mu$), written Erlang(k, μ), is the distribution of the sum of k i.i.d. Exp(μ) random variables:

$$T \sim \text{Erlang}(k, \mu) \Leftrightarrow f_T(t) = \frac{t^{k-1}\exp(-t/\mu)}{\mu^k (k-1)!}, \quad t \geq 0.$$

Note that Erlang$(1, \mu) \equiv$ Exp(μ).

A.2.7 Gamma distribution

The gamma distribution with shape parameter $k > 0$ and scale parameter $\mu > 0$, written Gamma(k, μ), is just the generalization of the Erlang(k, μ) distribution to the case where k is not an integer:

$$\Gamma \sim \text{Gamma}(k, \mu) \Leftrightarrow f_\Gamma(\gamma) = \frac{\gamma^{k-1}\exp(-\gamma/\mu)}{\mu^k \Gamma(k)}, \quad \gamma \geq 0,$$

where $\Gamma(\cdot)$ is the gamma function:

$$\Gamma(z) = \int_0^\infty t^{z-1}\, \mathrm{e}^{-t}\, \mathrm{d}t.$$

A.2.8 Nakagami distribution

The relation of the Nakagami distribution to the gamma distribution is the same as that of the Rayleigh distribution to the exponential distribution, namely if the power (i.e. square of the magnitude) of the complex baseband signal has the gamma distribution, the magnitude of the complex baseband signal is Nakagami distributed. More precisely, if $\Gamma \sim$ Gamma(k, μ), then $\sqrt{\Gamma} \sim$ Nakagami$(k, k\mu)$:

$$X \sim \text{Nakagami}(m, \omega) \Leftrightarrow f_X(x) = \frac{2m^m}{\Gamma(m)\omega^m} x^{2m-1} \exp\left(-\frac{m}{\omega}x^2\right),$$
$$x \geq 0,\ m > 0, \omega > 0.$$

Note that if $X \sim$ Nakagami$(k, k\mu)$ with $k \in \{1, 2, \dots\}$ and $\mu > 0$, then $X^2 \sim$ Erlang(k, μ). Also, Nakagami$(1, 2\sigma^2) \equiv$ Rayleigh(σ).

A.2.9 Lognormal distribution

If $X \sim \mathcal{N}(\mu, \sigma^2)$ for some μ and $\sigma > 0$, then $Y = \mathrm{e}^X$ is said to be lognormally distributed with parameters μ and σ^2, written $\mathcal{LN}(\mu, \sigma^2)$:

$$Y \sim \mathcal{LN}(\mu, \sigma^2) \Leftrightarrow f_Y(y) = \frac{1}{y\sqrt{2\pi\sigma^2}} \exp\left[-\frac{(\ln y - \mu)^2}{2\sigma^2}\right], \quad y > 0.$$

In the context of wireless communications, it is customary to use the exponent 10 instead of e and say that $X \sim \mathcal{LN}(\mu, \sigma^2)$ if $X_{\text{dB}} \equiv 10\log_{10} X \sim \mathcal{N}(\mu, \sigma^2)$, i.e. if $X \equiv \mathrm{e}^{X_{\text{dB}}(\ln 10)/10}$, where $X_{\text{dB}} \sim \mathcal{N}(\mu, \sigma^2)$. In this case, the units of μ and σ are also taken to be decibels (dB).

Appendix B HCNs in LTE

B.1 3GPP and LTE

The scale and complexity of design, manufacture, and deployment of wireless cellular systems make it all but impossible for a single-company proprietary architecture to gain traction in the global telecommunications market. Instead, cellular network operators and equipment vendors from all over the world have joined to form standards bodies at the national and international level in order to facilitate the evolution of cellular communication systems. At the present time, six national standards bodies (ARIB and TTC from Japan, CCSA from China, ATIS from North America, ETSI from Europe, and TTA from Korea) have combined to form a single international standards organization for wireless cellular communications, called the *Third Generation Partnership Project*, or 3GPP for short. The 3GPP partnership was first created to further the UMTS standard, which defined the so-called "third generation" of cellular communications using code division multiple access (CDMA) on the air-interface, but the name 3GPP was retained for the "fourth generation" using OFDMA. This fourth generation of cellular systems came to be known as the "long-term evolution" of the 3GPP standard, and is simply abbreviated as 3GPP-LTE, or just LTE for short.

B.2 Support for HCNs in LTE

3GPP standards are published as "Releases," with a typical interval of 12 to 18 months between releases. As the universal mobile telecommunications system (UMTS) standard became mature and the standards activity ramped up on LTE, the same release (Releases 8 and 9) contained standards specifications for both UMTS and LTE.

Release 9 provided end-to-end support for LTE "small cells," but these were assumed to be femtocells deployed in homes (called "home enhanced Node Bs" or HeNBs, "Node B" being the technical term for a BS in 3GPP), with backhaul provided by the internet service provider to the subscriber's home. However, there was no widespread deployment of such an LTE HeNB network. This standard also ported the "high interference indicator" flag from UMTS, whereby a cell suffering high interference on some orthogonal frequency division multiplexing (OFDM) subcarriers could broadcast this flag so that its neighboring cells could reduce their activity (or the activity of their served users) when they detected the flag. This frequency-domain information exchange was the only form of inter-cell interference coordination (ICIC) supported by this standard.

Release 10, from 2011, is taken to be the first release that is exclusively for LTE, and the specification of LTE in 3GPP Release 10 is called *LTE-Advanced*, or LTE-A for short. An excellent description of the LTE and LTE-A (Releases 8–10) standards is provided in Sesia, Toufik & Baker (2011). The Release 10 standard includes not only several enhancements to the macrocellular LTE standard, but also support for co-channel macro-pico HCNs, where the pico deployment is also owned and provisioned by the operator of the macrocellular tier. A key feature is the introduction of almost-blank subframes (ABSs) together with the (near-) zero power transmissions by the macro BSs during these ABSs in order to reduce interference to users (UEs, in 3GPP terminology) that have been induced to associate with pico BSs due to range expansion bias. This is called *enhanced ICIC*, or eICIC, to emphasize the evolution over the ICIC version in Release 9. Note that eICIC is a time-domain method.

Release 11, from 2012 (also called LTE-A), adds more capabilities to the eICIC version from Release 10, principally the support for enhanced algorithms at the receiver to minimize interference. This form of ICIC is called *further enhanced ICIC*, or feICIC for short. The use of reduced-power macro BS transmissions during ABSs was also debated but ultimately postponed to Release 12 (which is still being finalized and will probably be published in 2014). Since Release 12 follows up on the LTE-A standard from Release 11, it has been referred to in some quarters as LTE-B, but the 3GPP's official stance on naming is that, until further notice, all releases from Release 10 onward are to be referred to as LTE-A.

References

Abdulkafi, A., Kiong, T., Koh, J., Chieng, D. & Ting, A. (2012), "Energy efficiency of heterogeneous cellular networks: a review," *Journal of Applied Sciences* **12**(14), 1418–1431.

Ali, O. B. S., Cardinal, C. & Gagnon, F. (2010), "Performance of optimum combining in a Poisson field of interferers and Rayleigh fading channels," *IEEE Transactions on Wireless Communications* **9**(8), 2461–2467.

Aljuaid, M. & Yanikomeroglu, H. (2010), "Investigating the Gaussian convergence of the distribution of the aggregate interference power in large wireless networks," *IEEE Transactions on Vehicular Technology* **59**(9), 4418–4424.

Almhana, J., Liu, Z., Choulakian, V. & McGorman, R. (2006), "A recursive algorithm for gamma mixture models," in *Proc. IEEE Int. Conf. Communications (ICC)*, Istanbul, Jun. 11–15, 2006, pp. 197–202.

Andrews, J., Baccelli, F. & Ganti, R. (2011), "A tractable approach to coverage and rate in cellular networks," *IEEE Transactions on Communications* **59**(11), 3122–3134.

Andrews, J., Ganti, R., Haenggi, M., Jindal, N. & Weber, S. (2010), "A primer on spatial modeling and analysis in wireless networks," *IEEE Communications Magazine* **48**(11), 156–163.

Atapattu, S., Tellambura, C. & Jiang, H. (2011), 'A mixture gamma distribution to model the SNR of wireless channels', *IEEE Transactions on Wireless Communications* **10**(12), 4193–4203.

Auer, G., Giannini, V., Desset, C., Godor, I., Skillermark, P., Olsson, M., Imran, M., Sabella, D., Gonzalez, M., Blume, O. & Fehske, A. (2011a), "How much energy is needed to run a wireless network?," *IEEE Wireless Communications Magazine* **18**(5), 40–49.

Auer, G., Giannini, V., Godor, I., Skillermark, P., Olsson, M., Imran, M., Sabella, D., Gonzalez, M., Desset, C. & Blume, O. (2011b), "Cellular energy efficiency evaluation framework," in *IEEE 73rd Vehicular Technology Conf. (VTC Spring 2011)*, Budapest, May 15–18, 2011, pp. 1–6.

Baccelli, F. & Błaszczyszyn, B. (2009), *Stochastic Geometry and Wireless Networks, Volume 1: Theory*, Foundations and Trends in Networking. Hanover, MA: NOW Publishers Inc.

Baddeley, A. J. (2007), *Spatial Point Processes and their Applications*, Lecture Notes in Mathematics: Stochastic Geometry. Berlin: Springer Verlag.

Berman, A. & Plemmons, R. J. (1994), *Nonnegative Matrices in the Mathematical Sciences*. Philadelphia, PA: Society for Industrial and Applied Mathematics.

Błaszczyszyn, B., Karray, M. K. & Keeler, H. P. (2013), "Using Poisson processes to model lattice cellular networks," in *Proc. IEEE INFOCOM 2013*, Turin, Apr. 14–19, 2013, pp. 773–781.

Błaszczyszyn, B., Karray, M. K. & Klepper, F. X. (2010), "Impact of the geometry, path-loss exponent and random shadowing on the mean interference factor in wireless cellular networks," in *Proc. 3rd Joint IFIP Wireless and Mobile Networking Conf. (WMNC)*, Budapest. Oct. 13–15, 2010, pp. 1–6.

Bossmin, D. (2012), "Energy efficiency metrics: coverage and capacity together for good measure?," *1st ETSI Workshop on Energy Efficiency*, Genoa, Jun. 20–21, 2012, http://docbox.etsi.org/workshop/2012/201206_EEWORKSHOP/02_STANDARDIZATION_INITIATIVES/ATIS_BOSSMIN.pdf.

Caire, G., Adhikary, A. & Ntranos, V. (2013), "Physical layer techniques for cognitive femtocells," in T. Q. S. Quek, G. de la Roche, I. Güvenç & M. Kountouris, eds., *Small Cell Networks: Deployment, PHY Techniques, and Resource Management*. Cambridge: Cambridge University Press.

Calka, P. (2010), "Tessellations," in W. S. Kendall & I. Molchanov, eds., *New Perspectives in Stochastic Geometry*. Oxford: Oxford University Press.

Chen, L., Chen, W., Wang, B., Zhang, X., Chen, H. & Yang, D. (2011), "System-level simulation methodology and platform for mobile cellular systems," *IEEE Communications Magazine* **49**(7), 148–155.

Dahama, R., Sowerby, K. & Rowe, G. (2009), "Outage probability estimation for licensed systems in the presence of cognitive radio interference," in *IEEE 69th Vehicular Technology Conf. (VTC Spring 2009)*, Barcelona, Apr. 26–29, 2009, pp. 1–5.

DeVore, R. A. & Lorentz, G. G. (1993), *Constructive Approximation*. Berlin: Springer.

Dhillon, H., Ganti, R. & Andrews, J. (2013), "Modeling non-uniform UE distributions in downlink cellular networks," *IEEE Wireless Communications Letters* **2**(3), 339–342.

Dhillon, H., Ganti, R., Baccelli, F. & Andrews, J. (2011), "Coverage and ergodic rate in K-tier downlink heterogeneous cellular networks," in *Communication, Control, and Computing (Allerton), 2011 49th Annual Allerton Conf.*, Monticello, IL, Sept. 28–30, pp. 1627–1632.

Dhillon, H., Ganti, R., Baccelli, F. & Andrews, J. (2012), "Modeling and analysis of K-tier downlink heterogeneous cellular networks," *IEEE Journal on Selected Areas in Communications* **30**(3), 550–560.

Diggle, P. J. (2003), *Statistical Analysis of Spatial Point Patterns*, 2nd edn. London: Hodder Educational Publishers.

Erturk, M., Mukherjee, S., Ishii, H. & Arslan, H. (2013), "Distributions of transmit power and SINR in device-to-device networks," *IEEE Communications Letters* **17**(2), 273–276.

Fehske, A., Fettweis, G., Malmodin, J. & Biczok, G. (2011), "The global footprint of mobile communications: the ecological and economic perspective," *IEEE Communications Magazine* **49**(8), 55–62.

Foschini, G. & Miljanic, Z. (1993), "A simple distributed autonomous power control algorithm and its convergence," *IEEE Transactions on Vehicular Technology* **42**(4), 641–646.

Foss, S. & Zuyev, S. (1996), "On a certain Voronoi aggregative process related to a bivariate Poisson process," *Advances in Applied Probability* **28**, 965–981.

Friedland, S. & Krop, E. (2006), "Exact conditions for countable inclusion-exclusion identity and extensions," *International Journal of Pure and Applied Mathematics* **29**(2), 177–182.

Ge, X., Huang, K., Wang, C.-X., Hong, X. & Yang, X. (2011), "Capacity analysis of a multi-cell multi-antenna cooperative cellular network with co-channel interference," *IEEE Transactions on Wireless Communications* **10**(10), 3298–3309.

Gloaguen, C., Voss, F. & Schmidt, V. (2009), "Parametric distance distributions for fixed access network analysis and planning," in *21st Int. Teletraffic Congress (ITC 21 2009)*, Paris. Sept. 15–17, 2009, pp. 1–8.

Gradshteyn, I. S. & Ryzhik, I. M. (2000), *Table of Integrals, Series, and Products*, 6th edn. San Diego, CA: Academic Press.

Gulati, K., Evans, B. L., Andrews, J. G. & Tinsley, K. R. (2010), "Statistics of co-channel interference in a field of Poisson and Poisson-Poisson clustered interferers," *IEEE Transactions on Signal Processing* **58**(10), 6207–6222.

Haenggi, M. (2008), "A geometric interpretation of fading in wireless networks: theory and applications," *IEEE Transactions on Information Theory* **54**(12), 5500–5510.

Haenggi, M. (2012), *Stochastic Geometry for Wireless Networks*, Cambridge: Cambridge University Press.

Haenggi, M. & Ganti, R. K. (2009), *Interference in Large Wireless Networks*, Foundations and Trends in Networking. Hanover, MA: NOW Publishers Inc.

Haenggi, M., Andrews, J., Baccelli, F., Dousse, O. & Franceschetti, M. (2009), "Stochastic geometry and random graphs for the analysis and design of wireless networks," *IEEE Journal on Selected Areas in Communications* **27**(7), 1029–1046.

Heath, R. W., Kountouris, M., & Bai, T. (2013), "Modeling heterogeneous network interference using Poisson point processes," *IEEE Transactions on Signal Processing* **61**(16), 4114–4216.

Hunter, A. M., Andrews, J. G. & Weber, S. (2008), "Transmission capacity of ad hoc networks with spatial diversity," *IEEE Transactions on Wireless Communications* **7**(12), 5058–5071.

Ito, K., ed. (1993), *Encyclopedic Dictionary of Mathematics*, 2nd edn. Cambridge, MA: MIT Press.

Jacquet, P. (2008), *Realistic Wireless Network Model with Explicit Capacity Evaluation*, Tech. Rep. RR-6407. Rocquencourt: INRIA.

Jo, H. S., Sang, Y. J., Xia, P. & Andrews, J. G. (2012), "Heterogeneous cellular networks with flexible cell association: a comprehensive downlink SINR analysis," *IEEE Transactions on Wireless Communications* **11**(10), 3484–3494.

Kelly, F. P. (2011), *Reversibility and Stochastic Networks*, rev. edn. Cambridge: Cambridge University Press.

Kingman, J. F. C. (1993), *Poisson Processes*. Oxford: Oxford University Press.

Kleptsyna, M. L., Le Breton, A. & Viot, M. (2002), "General formulas concerning Laplace transforms of quadratic forms for general Gaussian sequences," *Journal of Applied Mathematics and Stochastic Analysis* **15**(4), 309–325.

Lapidoth, A. (1996), "Nearest neighbor decoding for additive non-Gaussian noise channels," *IEEE Transactions on Information Theory* **42**(5), 1520–1529.

Liu, E., Zhang, Q. & Leung, K. (2011), "Asymptotic analysis of proportionally fair scheduling in Rayleigh fading," *IEEE Transactions on Wireless Communications* **10**(6), 1764–1775.

Madhusudhanan, P., Restrepo, J. G., Liu, Y. E. & Brown, T. X. (2009), "Carrier to interference ratio analysis for the shotgun cellular system," in *Proc. IEEE Global Communications Conf. (Globecom) 2009*, Honolulu, HI, Nov. 30–Dec. 4, 2009, pp. 1–6.

Madhusudhanan, P., Restrepo, J. G., Liu, Y. E., Brown, T. X. & Baker, K. R. (2011), "Multi-tier network performance analysis using a shotgun cellular system," in *Proc. IEEE Global Communications Conf. (Globecom) 2011*, Houston, TX, Dec. 5–11, 2011, pp. 1–6.

Madhusudhanan, P., Restrepo, J. G., Liu, Y. E. & Brown, T. X. (2012a), "Downlink coverage analysis in a heterogeneous cellular network," in *Proc. IEEE Global Communications Conf. (Globecom) 2012*, Anaheim, CA. Dec. 3–7, 2012, pp. 4170–4175.

Madhusudhanan, P., Restrepo, J. G., Liu, Y. E., Brown, T. X. & Baker, K. R. (2012b), "Stochastic ordering based carrier-to-interference ratio analysis for the shotgun cellular system," *IEEE Wireless Communications Letters* **1**(6), 565–568.

Mallik, R. (2010), "A new statistical model of the complex Nakagami-m fading gain," *IEEE Transactions on Communications* **58**(9), 2611–2620.

Møller, J. & Waagepetersen, R. P. (2004), *Statistical Inference and Simulation for Spatial Point Processes*. Boca Raton, FL: Chapman & Hall/CRC.

Mukherjee, S. (2011a), "Downlink SINR distribution in a heterogeneous cellular wireless network with max-SINR connectivity," in *Proc. 2011 Allerton Conf.*, Monticello, IL, Sept. 28–30, 2011, pp. 1649–1656.

Mukherjee, S. (2011b), "UE coverage in LTE macro network with mixed CSG and open access femto overlay," in *Proc. IEEE Int. Conf. Communications Workshops (ICC), 2011*, Kyoto, Jun. 5, 2011, pp. 1–6.

Mukherjee, S. (2012a), "Downlink SINR distribution in a heterogeneous cellular wireless network," *IEEE Journal on Selected Areas in Communications* **30**(3), 575–585.

Mukherjee, S. (2012b), "Downlink SINR distribution in a heterogeneous cellular wireless network with biased cell association," in *IEEE Int. Conf. Communications (ICC) 2012, Small Cells Workshop (SmallNets)*, Ottawa, Jun. 15, 2012, pp. 6780–6786.

Mukherjee, S. (2013), "Coverage analysis using the Poisson point process model in heterogeneous networks," in T. Q. S. Quek, G. de la Roche, I. Güvenç & M. Kountouris, eds., *Small Cell Networks: Deployment, PHY Techniques, and Resource Management*. Cambridge: Cambridge University Press.

Mukherjee, S. & Güvenç, I. (2011), "Effects of range expansion and interference coordination on capacity and fairness in heterogeneous networks," in *Signals, Systems and Computers (ASILOMAR), 2011 Conf. Record 45th Asilomar Conf.*, Pacific Grove, CA, Nov. 6–9, 2011, pp. 1855–1859.

Mukherjee, S. & Ishii, H. (2013), "Energy efficiency in the phantom cell enhanced local area architecture," in *2013 IEEE Wireless Communications and Networking Conf.*, Shanghai, Apr. 7–10, 2013, pp. 1267–1272.

NTT DoCoMo (2007), "Discussions on UE capability for dual-antenna receiver," 3GPP Standard Contribution (R4-070745), in *TSG-RAN Working Group 4 Meeting #43*, Kobe, May 7–11, 2007, agenda item 6.4.2.

Papoulis, A. (1991), *Probability, Random Variables, and Stochastic Processes*, 3rd edn. New York: McGraw-Hill.

Pearson, J. (2009), "Computation of hypergeometric functions." Unpublished Master's thesis, Worcester College, University of Oxford.

Pinto, P. C. & Win, M. Z. (2010a), "Communication in a Poisson field of interferers – Part I: Interference distribution and error probability," *IEEE Transactions on Wireless Communications* **9**(7), 2176–2186.

Pinto, P. C. & Win, M. Z. (2010b), "Communication in a Poisson field of interferers–Part II: Channel capacity and interference spectrum," *IEEE Transactions on Wireless Communications* **9**(7), 2187–2195.

Schrijver, A. (1986), *Theory of Linear and Integer Programming*. Chichester: J. Wiley.

Sesia, S., Toufik, I. & Baker, M., eds. (2011), *LTE, the UMTS Long Term Evolution: From Theory to Practice*, 2nd edn. Chichester: Wiley.

Streit, R. L. (2010), *Poisson Point Processes – Imaging, Tracking, and Sensing*. New York: Springer Verlag.

Suryaprakash, V., dos Santos, A. F., Fehske, A. & Fettweis, G. P. (2012), "Energy consumption analysis of wireless networks using stochastic deployment models," in *2012 IEEE Global Communications Conf. (GLOBECOM)*, Anaheim, CA: Dec. 3–7, 2012, pp. 3177–3182.

3GPP (2010), *Evolved Universal Terrestrial Radio Access (E-UTRA); User Equipment (UE) Procedures in Idle Mode*, TS 36.304, 3rd Generation Partnership Project (3GPP).

Thummler, A., Buchholz, P. & Telek, M. (2006), "A novel approach for phase-type fitting with the EM algorithm," *IEEE Transactions on Dependable and Secure Computing* **3**(3), 245–258.

Torrieri, D. & Valenti, M. (2012), "The outage probability of a finite ad hoc network in Nakagami fading," *IEEE Transactions on Communications* **60**(11), 3509–3518.

Voss, F., Gloaguen, C., Fleischer, F. & Schmidt, V. (2009), "Distributional properties of euclidean distances in wireless networks involving road systems," *IEEE Journal on Selected Areas in Communications* **27**(7), 1047–1055.

Vu, T. T., Decreusefond, L. & Martins, P. (2012), "An analytical model for evaluating outage and handover probability of cellular wireless networks," in *2012 15th International Symposium on Wireless Personal Multimedia Communications (WPMC)*, Taipei, Sept. 24–27, 2012. pp. 643–647.

Weber, S. & Andrews, J. G. (2012), *Transmission Capacity of Wireless Networks*, Foundations and Trends in Networking. Hanover, MA: NOW Publishers Inc.

Williams, D. (2001), *Weighing the Odds: A Course in Probability and Statistics*. Cambridge: Cambridge University Press.

Yu, B., Mukherjee, S., Ishii, H. & Yang, L. (2012), "Dynamic TDD support in the LTE-B enhanced local area architecture," in *2012 IEEE Globecom Workshops (GC Wkshps)*, Anaheim, CA, Dec. 3–7, 2012, pp. 585–591.

Yu, S. & Kim, S.-L. (2013), "Downlink capacity and base station density in cellular networks," in *11th Int. Symp. Modeling and Optimization in Mobile, Ad Hoc and Wireless Networks (WiOpt)*, Tsukuba Science City, May 13–17, 2013, pp. 119–124.

Zhang, X. & Haenggi, M. (2012), "Random power control in Poisson networks," *IEEE Transactions on Communications* **60**(9), 2602–2611.

Zuyev, S. (2010), "Stochastic geometry and telecommunications networks," in W. S. Kendall & I. Molchanov, eds., *New Perspectives in Stochastic Geometry*. Oxford: Oxford University Press.

Author index

Subject index